THE AI PARADOX

The AI Paradox

HOW TO MAKE SENSE
OF A COMPLEX FUTURE

VIRGINIA DIGNUM

PRINCETON UNIVERSITY PRESS

PRINCETON & OXFORD

Published by Princeton University Press
41 William Street, Princeton, New Jersey 08540
99 Banbury Road, Oxford OX2 6JX

press.princeton.edu

GPSR Authorized Representative: Easy Access System Europe -
Mustamäe tee 50, 10621 Tallinn, Estonia, gpsr.requests@easproject.com

All Rights Reserved

ISBN 9780691269085
ISBN (epub) 9780691287270
ISBN (PDF) 9780691269078
Library of Congress Control Number: 2025944766

British Library Cataloging-in-Publication Data is available

Editorial: Hallie Stebbins and Chloe Coy
Production Editorial: Terri O'Prey
Jacket/Cover Design: Karl Spurzem
Production: Erin Suydam
Publicity: Maria Whelan and Kate Farquhar-Thomson
Copyeditor: Madeleine B. Adams

This book has been composed in Arno

10 9 8 7 6 5 4 3 2 1

CONTENTS

PARADOXES HAVE always fascinated me. They challenge our assumptions, force us to think deeper, and reveal that reality is rarely as simple as it seems. This book was born from the realization that much of what we believe to be true is shaped by how we frame a question, not just by the answer itself.

For several decades, I have worked in the field of artificial intelligence (AI), studying its responsible development, impact, and regulation. As an international expert deeply engaged in AI governance, my focus has always been on ensuring that AI serves humanity, rather than the other way around. AI is reshaping our world; its promises are vast but so are its risks. The way we design, use, and regulate AI will determine whether it amplifies human potential or deepens societal inequalities. And yet, AI itself is full of paradoxes.

One of these paradoxes is how something once dismissed as a niche curiosity is now seen as the force that will redefine our future. Early in my career, while working in the automotive industry, I developed an AI system to optimize the assembly process of vehicles. One day, a senior manager pulled me aside and said, "This system is fantastic; it's saving us time and money. But please, never tell anyone outside your group that we are using artificial intelligence. We don't want to be associated with such a weird technology." At the time, AI was met with skepticism and often dismissed as impractical or even irrelevant. Decades later, the conversation has reversed. Now, everyone wants to claim they were ahead of the AI revolution. But another paradox remains: as AI becomes more powerful, our understanding of its role in society becomes more complex. My journey, from developing early expert systems in the mid-1980s to becoming a professor in responsible AI and advising global institutions, has shown me that technology is never neutral. The way we choose to design, use, and regulate AI will determine whether it truly benefits humanity or deepens existing divides.

The idea of paradox can be understood in two ways: a logical paradox presents an inherent contradiction that defies logic, whereas a paradoxical truth only appears contradictory but reveals a deeper reality upon closer examination. An example of a logical paradox is: *"This statement is false."* If it's true, then it's false. But if it's false, then it must be true. A paradoxical truth is exemplified by the sentence *"Less is more,"* meaning that simplicity can be more effective than excess. The paradoxes explored in this book may not always fit the formal definition of a logical paradox, but that's precisely the point. My aim is to stimulate reflection, to show that every concept, every truth, every certainty has another side. By engaging with contradictions, we learn to question our perspective, sharpen our reasoning, and, most important, form our own opinions rather than blindly adopting those of others.

Paradoxically, the more AI can do, the more it highlights what makes human intelligence unique. The more we rely on AI, the less we seem to question its impact. The more we debate AI's risks, the harder it becomes to agree on what AI even is. These contradictions are not just intellectual curiosities; they are essential to understanding the technology shaping our future. But paradoxes extend far beyond AI. They exist in justice, power, regulation, and intelligence itself. Exploring them helps us see different sides of an issue, question dominant narratives, and, ultimately, form our own opinions instead of being led by others.

We live in an age of information overload, where it's easier than ever to let others do the thinking for us. But true understanding requires effort. It means grappling with complexity, sitting with discomfort, and seeing beyond the obvious. Paradoxes remind us that certainty is elusive and that wisdom often lies in embracing contradiction.

This book is an invitation to think critically about the world we are building. I invite you to explore the paradoxes not as a collection of riddles to be solved, but as a guide to seeing complexity with clarity, questioning what seems inevitable, and engaging in the choices that shape our shared future.

THE AI PARADOX

1

The AI Paradox

THE IRREPLACEABLE HUMAN

ARTIFICIAL INTELLIGENCE (AI) is shaping our world in ways big and small, from helping us find the fastest route home to sparking debates about whether it could one day replace human workers or even outthink us altogether. Just think of the maps app on your phone. It quickly calculates the best route based on real-time traffic, something that would take a human much longer to figure out. Yet, if the app suggests a route through a dangerous area, it doesn't understand the broader context or consequences of sending you in that direction; it just follows the data about road and traffic conditions. This mix of impressive capabilities and clear limitations is what makes AI both exciting and deeply complex. As it evolves, it challenges us to consider not just what it can do but also what it cannot, and whether it could ever truly replicate the depth of human intelligence.

Throughout this book, we'll explore why, despite AI's potential to mimic or outperform certain abilities, the distinct qualities of human intelligence, such as our emotional depth, creativity, ethical insight, and capacity for complex reasoning, remain beyond the reach of any machine. We will see that these uniquely human traits, which are woven together in intricate and evolving ways, cannot be fully replaced, no matter how advanced AI becomes. For instance, although AI can quickly analyze vast datasets, provide advice, generate text, and offer diagnostic insights, it cannot grasp the nuanced emotional cues in a conversation or generate or appreciate genuinely innovative artistic expressions.[1] For instance, Emily M. Bender, a linguist at the University of Washington, and colleagues, emphasize that large language models like Chat-GPT replicate language patterns without true comprehension, leading to potential inaccuracies and a lack of genuine empathy.[2] Similarly, Shannon Vallor, a philosopher of technology, points out that although AI can mimic human traits, it lacks the capacity for virtues like courage, honesty, and empathy, which are fundamental to human experience. She warns that although AI might simulate emotions, it doesn't possess the genuine emotional depth that characterizes human interactions.[3]

Humans create timeless art, write evocative literature, and compose soul-stirring music, forms of creativity that AI can imitate but most likely not authentically replicate, raising the question of whether intelligence lies in the process, the outcome, or a combination of both. Additionally, humans possess moral and ethical discernment, that is, the ability to distinguish right from wrong, which allows us to navigate social complexities and make decisions based on empathy, values, and cultural context. This multidimensional form of intelligence enables humans to excel in areas like negotiation, leadership, and caregiving, where emotional intelligence and ethical considerations are crucial. In fact, the uniqueness of human intelligence is what enables the development of AI and its evolving capabilities in the first place. It is the human capacity for abstract think-ing, experimentation, and ethical reflection that drives the development and refinement of AI technologies. Thus, even as AI continues to evolve, the depth and breadth of human thought and experience remain unparalleled and foundational to technological advancement, as we will discuss in chapter 3. The fundamental paradox of AI is that, even in a machine-dominated era, human intelligence is still crucial.

The AI Paradox
The more AI can do, the more it highlights the irreplaceable nature of human intelligence.

Given this, on the one hand, it may be reassuring to know that we people will not be soon, or ever, replaced or taken over by AI; on the other hand, we do need to ask ourselves some fundamental and hard questions:

- How do we ensure safety and control over AI to prevent it from being used against human interests, including mitigating existential risks?
- Will all people benefit equally and equitably from the advances of AI technology, or will some of us be more "replaceable" than others? If so, who is replaceable and who will stay ahead?
- Will AI development contribute to even larger power imbalances and economic disruptions, such as widespread unemployment? How can we ensure that everybody can realize their full potential and unique capabilities?
- How can we address the unintended consequences that may arise from superintelligence? In particular, what

measures can prevent the weaponizing and misuse of AI for malicious purposes?

• Is there a risk that overreliance on AI could diminish human autonomy and critical thinking skills?

This book will explore these pressing questions one by one, uncovering the paradoxes at the core of each. Paradoxes challenge our assumptions, force us to think more deeply and reveal that reality is rarely as simple as it seems, leading us to realize that much of what we believe to be true is shaped by how we frame a question, not just by the answer itself. Central to the book is the core AI paradox, which we will explore in this chapter. It will help us get a better grip on AI's potential, its limits, and the key choices we face about its role in our lives. It is up to us to engage with these challenges, embrace the opportunities, and guide AI development toward a future that reflects our values and aspirations.

AI is challenging to define, but that does not hinder our ability to talk about it and to conceptualize it. In fact, we are all quite capable of engaging in meaningful discussions about it without being constrained by the lack of a formal definition. Our understanding of AI is shaped by its real or perceived capabilities, while we also continuously shape it through our conceptualizations and narratives. But because we need to start somewhere, a commonly accepted definition of AI comes from the OECD:[4] *"An AI system is a machine-based system that, for explicit or implicit objectives, infers, from the input it receives, how to generate outputs such as predictions, content, recommendations, or decisions that can influence physical or virtual environments. Different AI systems vary in their levels of autonomy and adaptiveness after deployment."*[5] Interestingly, this definition, updated in 2023, refines their original 2018 definition, which highlights the inherent challenge of defining AI.[6] We will explore this issue further in chapter 2.

Independently of how we define it, for many of us, AI may feel like the weather, something that happens to us but is beyond our control. We adapt to the weather by carrying an umbrella when rain is expected or planning a picnic when the forecast is sunny. Yet AI is not the weather. Even though its behavior can sometimes seem unpredictable or difficult to understand, it is entirely a human creation. AI is more like a car engine. It is built by people, following specific instructions to achieve certain goals and meet particular requirements. It is a product of human creativity, shaped by our choices and intentions. Which means that we are not powerless. We have the ability to decide how AI is designed, developed, and applied to shape the outcomes we want. Ultimately, the results depend on the decisions we, as people, make.

This does not mean that we all have to become experts on the technology and the methods that make AI systems possible. For one, I don't really understand how a car engine works, nor am I able to build or repair a car engine. But I understand it sufficiently to know how to drive a car, and how to interact with other traffic on the road. In the same way, we need to be able to have some understanding of what AI can do, what it is and why it is developed, for whom and by whom it is developed, and what AI can do for us. In chapter 2 we will further discuss what AI is and what options we have.

For better or worse, AI is being defined by us, people. Then, shouldn't we all have something to say about it? To be part of this discussion, to be able to demand means by which to govern AI, to demand accountability for what is done using AI, and to define the boundaries of when and where AI can and should be used, means that we all need to be able to understand what AI can and cannot do, and what its consequences are.

By accepting that AI "happens to us," we are accepting that others will determine what AI will do to all of us. As we will see in chapter 6, there are now some who benefit from this situation: the tiny group that currently has the power to decide about how to develop and use AI. But we cannot accept that we have no say on how it interacts with us, or passively let it be "dumped" on us. It is easy to let it be, but now is the time to determine how it should be. If we again take the automotive industry as an example, the continuous improvements we see in cars are not only the result of industry choices but are also guided by public opinion and regulation. Our responsibility starts with our ability and our willingness to be well informed and to contribute to the discussion; to use our voices as informed and concerned citizens of the world.

I've been working in the field of AI, both in industry and in academia, since the late 1980s. I developed my first AI system in 1986, an expert system to determine eligibility for social housing. Since then, I've not only witnessed but also directly experienced the field's ups and downs. Never before has there been such widespread excitement, and fear, across so many sectors as we have seen in the past decade, especially since the launch of large language models (LLMs) and other generative AI applications at the end of 2022.[7] The true and sustainable future of AI depends on our recognition that its potential is fundamentally rooted in human intelligence. I understand AI deeply and know how to build these systems, but the more I learn, the more I see that AI can never replace the essential ingenuity and creativity that only humans possess.

Take, for example, the use of AI in medical imagery. AI algorithms, particularly those based on machine learning, are increasingly used to analyze medical images like X-rays, MRIs, and CT scans. Sometimes algorithms can detect patterns and anomalies with greater speed and accuracy than human radiologists. This led, some years ago, several experts, including 2024 Nobel Prize winner Geoff Hinton and the influential computer scientist Andrew Ng, to conclude that the job of radiologist is at risk of being replaced by AI. But they retracted that conclusion some years later, even though AI is very valuable to handle large volumes of diagnostic data, aiding in early detection of diseases such as cancer, which can significantly improve treatment outcomes. This advanced capability of AI, however, also highlights the indispensable role of human healthcare professionals.[8] Although AI can identify patterns and suggest diagnoses, it lacks the ability to consider not only the patient's full medical history but also their social context, lifestyle factors, and personal preferences.[9] AI also cannot provide the empathetic care and communication trust that are critical in the patient-doctor relationship.[10] So, even if AI platforms such as Google Health[11] or PathAI,[12] have demonstrated capabilities that can match or even exceed human performance in specific tasks, they are primarily designed to assist on very specific tasks and are not capable of fully replacing medical professionals. Doctors are essential for interpreting AI findings in the broader context of patient care and discussing treatment options with patients. Healthcare professionals are also responsible for the final diagnostic decisions, and are accountable for these. In fact, even if we think that machine-generated diagnostics may some day be more accurate than those of a medical professional, none of us would comfortably accept *"the computer said so"* as a justification for a medical decision, in particular if the computer cannot explain why. Human expertise and capability to take responsibility remain essential.

AI is often seen as more rational and objective than humans, leading to the belief that it will be able to make *"better"* decisions, and therefore soon replace us in many areas, including our workplaces. AI models interpret the world differently from humans. Current AI systems, particularly machine learning, excel at identifying patterns in data. For example, again in medical imaging, AI detects tumors by recognizing statistical correlations in images. In contrast, humans learn through relationships between concepts, often based on cause and effect. A doctor, for instance, relates lung cancer to symptoms like cough and shortness of breath, informed by a broader understanding of the patient's health, lifestyle, and context, elements that machine learning systems lack. AI models do not possess an ontology of the world or the ability to reason about relationships between concepts.

AI systems are designed to optimize results based on data-driven priorities, whereas human reasoning is rooted in understanding connections and meanings within a broader context. We don't just recognize patterns; we interpret them based on our knowledge of the world. Human cognition also involves abstract thinking, reasoning with limited data, and making inferences from incomplete patterns, skills that statistical methods used by AI cannot replicate, even with larger datasets. This leads to machine learning systems that encode the patterns and relationships in their training data but do not understand the objects those patterns represent. Their form of representation, though alien to us, can be valuable. For example, a system I once built for a national migration agency identified a seemingly bizarre pattern: a correlation between complex asylum cases and people born in January. The reason was simple: asylum seekers without documentation, who are most of the complex cases, are often assigned a "random" birth date, typically

January 1st. This pattern didn't imply causality but helped offi-
cials better understand the context of asylum applications. This
example shows that correlations are valuable for identifying
potential relationships worthy of further investigation, but do
not, on their own, provide a basis for formal causal inference
or for abstraction. On the whole, we have more to gain from
the combined abilities of humans and machines. As we will discuss in chapter 3, human and artificial intelligence not only are
different but also serve different purposes.

So, in a nutshell, it can be said that human reasoning uses
abstraction and causation as basic processes whereas machines
learn by correlation.[13] In *The Book of Why*,[14] computer scientist and philosopher Judea Pearl explains that humans are able
to ask "what if" questions, imagine different scenarios, and reason beyond just recognizing the correlations in data that AI and
machine learning rely on for finding patterns. AI, on the other
hand, can't create such models; it processes the information
it has seen without understanding the reasons behind events.
Abstraction and causation, which are key to human thinking, go
beyond what correlation alone can reveal. Seen as an evolutionary characteristic of humans,[15] abstraction enables us to survive
in a complex and dynamic world. Thinking in causal relations
allows us to make predictions on how it will evolve and how
our actions will change it. Those abstractions and models are
not 100 percent correct but are good enough for us to act on
them. Whereas correlations help identify relationships, causality requires human interpretation and reasoning. Correlations
can suggest associations but do not establish cause and effect.
AI can help identify patterns, but humans are needed for causal
inference and abstraction. The true potential lies in combining
AI and human intelligence to enhance our abilities rather than
replace them.

Human decision-making is often guided by intuition or a "gut feeling," a subconscious process machines cannot replicate. This instinctive aspect of human thought is crucial in everything from everyday choices to major life decisions, offering a depth of understanding that AI cannot achieve. Likewise, traits like empathy, self-awareness, and emotional intelligence are uniquely human and cannot be authentically emulated by AI. Although AI can mimic empathetic behavior, especially in therapeutic contexts, it lacks the genuine emotional experience inherent in humans. We humans also possess the ability for moral reasoning, making decisions based on ethical principles about what ought to be, whereas AI systems, driven by utility maximization and data-based logic, cannot distinguish the possible from the impossible or the useful from the useless—leading to phenomena like the "hallucinations" of large language models. Human intelligence is marked by flexibility and adaptability, allowing us to learn from limited information and apply knowledge across various contexts, an essential quality in an age where AI tackles increasingly complex tasks.

Our ability to operate across diverse contexts and our sense of empathy enable us to act not just for personal gain but out of selflessness, equity, and justice—even at the cost of our own interests. Guided by a sophisticated moral compass, we seek to prevent remorse and correct wrongs, aware of the impact our actions have on both ourselves and others. This moral dimension often leads us to "satisficing"—seeking solutions that are "good enough" in light of life's complexities. We live in a world full of shades of meaning, relational dynamics, and ethical ambiguities. The core difference lies not just in capabilities, but in the essence of being: AI calculates, while humans feel; AI iterates, while humans imagine.

A last important feature of human reasoning is that we do not glue social behavior on top of some intelligence. We behave socially before we behave intelligently. Our evolutionary history reveals that social behaviors like cooperation, communication, and group living were not just important for survival—they were the very foundation upon which our intelligence developed. As Frank Dignum, a well-known expert in social AI (and my husband), says: "*Intelligence is deeply grounded in our social nature; it's not merely that social abilities were added to a pre-existing intelligence, but rather that these abilities are the core of all our intelligence.*" From infancy, our cognitive growth is intimately linked to social interactions, as our brains are wired to understand others, navigate social relationships, and work within groups. Our need to belong, recognize social cues, and cooperate with others has shaped our cognitive evolution. Thus, our intelligence is a consequence of our sociality, making social abilities the essence of what it means to be intelligent. In contrast, AI systems are more akin to "lone wolves." AI systems are designed to process information, solve problems, or perform specific tasks independent of social context. Although AI systems can mimic social interactions or collaborate with humans, these abilities are typically layered onto a framework that was not inherently designed for social engagement. Sociality is, at best, a capability that developers attempt to build on top of their existing functionalities. This difference highlights a significant distinction: for humans, social intelligence is intrinsic, whereas for AI, it is an extrinsic feature that requires deliberate engineering.

As AI progresses, we will increasingly value the unique human abilities of imagining and creating, not solely based on existing data or patterns but fundamentally emerging from our social traits. For instance, the 2023 Hollywood writers' strike sought to establish regulations on the use of AI in content creation.[16] It turned out to be one of the longest strikes in that industry and underscored the tension between AI's analytical prowess and its lack of nuance in understanding context. This event was not merely a standoff to limit the capabilities of AI, and thereby a lost race, but a plea to use and develop AI differently. Stemming from concerns about unregulated AI usage potentially replacing human creativity and jobs in film and TV, the strike was a call for a balanced symbiosis between human ingenuity and machine efficiency. The outcome ensures that writers maintain control over how and when to use AI tools, and that studios cannot use AI or digital replications without the informed consent of the performer, thus preserving human roles in the creative process. This set a vital precedent for the industry, emphasizing the need to use AI as an augmentative tool rather than a replacement, thus enriching the creative landscape rather than impoverishing it. The strikes also challenged Hollywood power structures, securing significant gains for actors and writers. The paradox here is that by combating the use of AI in the industry, the strikes ultimately bolstered human creativity, underscoring the unique value of human imagination and securing greater rights and control for writers and performers in an increasingly AI-driven landscape.

The questions of what AI is, and how its capabilities differ from those of people and human organizations, become particularly relevant when we aim at governing or regulating it, as we will discuss in detail in chapter 5. What are we attempting to govern? What is the difference between AI and any other digital system? Is it the large amounts of data AI uses? But many other systems use big data, too. Is it the lack of transparency in AI? Even human organizations lack transparency. Or is our concern rooted in the fact that AI is often in the hands

of big private companies, which might operate beyond democratic oversight? Although market mechanisms are designed
to handle the power of large corporations, AI presents unique
challenges, in particular where it concerns the ability to act
autonomously, bringing with it questions about accountability
and unpredictability. AI systems, though efficient, don't have
the moral and ethical reasoning humans do. They operate on
algorithms and patterns, and lack the nuanced understanding humans bring to decision-making. This difference leads
to public debates, regulatory efforts, and the formation of
commissions and expert groups. Are we, perhaps, missing
something else in this conversation? The essence of these concerns seems to go beyond just the technical aspects of AI and
into the realm of its broader impact on society, ethics, and
governance.

So far so good. The reflections above seem to suggest that
although AI can replicate human intelligence in tasks like data
analysis and decision-making, it cannot replace the essential
human skills of critical thinking, empathy, and adaptability. In
fact, throughout this book, I emphasize the idea that, although
AI can process data and identify patterns at incredible speeds, it
operates within deterministic frameworks, lacking the human
abilities of intuition, moral judgment, and creativity. This
inherent limitation reinforces the idea that AI should be seen
as a complementary tool to human intelligence, not a replacement, as it cannot grasp the full depth of human reasoning,
context, and ethical decision-making, abilities that are crucial
for dealing with complex issues and maintaining meaningful
human connections, especially in fields like customer service,
healthcare, and creative industries. In these areas, the unique
touch of human interaction remains irreplaceable by machines.
Humans also play a pivotal role in creating AI itself, in tasks

like designing algorithms, curating and labeling data, setting objectives, and ensuring ethical standards. Currently, the complexity and creativity required in these tasks make it unlikely that AI can fully replace humans in AI development. As such, human oversight remains crucial for guiding AI toward beneficial and ethical outcomes, making the human role in AI creation irreplaceable at this stage.

This viewpoint raises an important question: Can it be sustained in the long term, especially when considering the potential rise of superintelligence? Superintelligence, defined as AI systems that surpass human intellect in nearly every domain, could fundamentally change our understanding of human intelligence's role. The advent of such AI systems, following the development of artificial general intelligence (AGI)[17] that operates at the level of human intelligence, would present unprecedented challenges and force us to reconsider the relevance and uniqueness of human cognitive abilities in a world where AI may exceed those capabilities.

The current debate is marked by a division between two camps: On one side, proponents of rapid AI development argue that it holds the key to solving global challenges such as climate change and healthcare, while driving economic growth and geopolitical advantage. They emphasize the importance of leading in AI to maintain a competitive edge and advocate for incremental improvements based on real-world deployment, seeing risks as overstated and manageable through innovation and adaptation. On the other side, those who prioritize caution focus on the existential risks advanced AI could pose to humanity, including the loss of control over AI systems. They highlight ethical concerns like bias, job displacement, and privacy issues, calling for strong governance and global cooperation. This camp stresses the need for precautions, believing that once certain advancements are made, they may be irreversible, making it crucial to slow progress and ensure AI is developed safely and ethically.

Critics of existential risk arguments contend that these concerns are speculative and overestimate AI's future capabilities, focusing on unlikely worst-case scenarios. At the same time, they argue that this debate has distracted from pressing issues such as bias, privacy, and job displacement. Focusing on distant, improbable threats causes resources to be misallocated, developments to remain out of the public eye, and real-world challenges to be left unaddressed. As we will explore in later chapters, focusing on current risks and challenges will help us develop AI in line with principles of trustworthiness, responsibility, and governance, which will make catastrophic outcomes less likely.

In my view, the quest for superintelligence or AGI is not just a triumph of technology but, in many ways, a failure of governance and common sense. The real threat comes not from the power of superintelligent systems but rather from our inability to use technology responsibly. Solving our complex societal problems is not about perfect technology, but about using technology alongside better governance, deeper reflection, and greater participation. When it comes to the complex, wicked problems we face today, such as climate change, migration, and democracy—there are no perfect answers.[18] Every solution comes with trade-offs, and the goal is to understand what is at stake and the consequences of each proposed solution, rather than accepting the "perfect" answer an AGI system may offer. The risk lies not in the specific decisions AI will make, but in granting AI the power to make decisions in the first place. We all have a role to play—one that machines can never replace—because we are an integral part of society and part of both the problem and the solution. In this book, we will explore the many options available to address these challenges. We have choices, and fortunately, not all of them rely on technology. But we must have the courage to make those choices and act on our ability to decide.

The Usefulness of Paradoxes

Paradoxes teach us to question and critically evaluate our assumptions, leading to a deeper understanding of the situations we encounter. Paradoxes highlight the complexity of the world around us. They remind us that simple, one-size-fits-all solutions are often inadequate for addressing complex problems, which require a more nuanced and contextual approach. In untangling this fundamental AI paradox we will encounter many more paradoxes. This book explores how these paradoxes can help us get a more informed understanding of the field, the vested interests and stakes involved, the opportunities and risks for individuals and society, and the implications for a sustainable and just future for all. I hope that it can help us get a deeper comprehension of our human role in an AI-mediated world, emphasizing our increasing responsibility for the technology we're creating and using to shape our world. The chapters of this book are arranged around the following paradoxes, challenging our intuitive assumptions about how AI works and what role it has in our society.

1. The AI Paradox: The more AI can do, the more it highlights the irreplaceable nature of human intelligence. (this chapter)
2. The Agreement Paradox: The more we explore AI, the harder it becomes to agree on its definition. (chapter 2)

3. The Intelligence Paradox: AI is what AI cannot do. (chapter 3)
4. The Justice Paradox: Less bias is not always more justice. (chapter 4)
5. The Regulation Paradox: Responsible innovation needs regulation. (chapter 5)
6. The Power Paradox: The more AI you get, the less control you have. (chapter 6)
7. The Superintelligence Paradox: The more we chase AGI, the more we discover that true superintelligence lies in human cooperation. (chapter 7)
8. The Solution Paradox: Solving problems with technology often creates more problems. (chapter 8)

Life is full of paradoxes, and they are not just silly or laughable; they often reveal flaws in how we understand a concept or situation. One famous example is Zeno's paradox: to reach a goal, you must first cover half the distance, then half of the remaining distance, and so on, suggesting you would never actually reach your destination. In everyday life, this reflects how breaking a larger goal into smaller tasks can sometimes feel overwhelming, preventing us from starting. Yet, dividing tasks into manageable steps is what allows us to make progress. Zeno's paradox teaches that overthinking can lead to inaction, and taking the first step is often better than being paralyzed by analysis.

This paradox also offers valuable insights for AI development. It aligns with the Intelligence Paradox discussed in chapter 3, where new challenges always arise just as we think we've achieved something. Instead of discouraging us, this highlights the importance of every incremental step in technological progress. Breakthroughs come from consistent effort over time—neural networks in the 1970s, natural language processing in the 1950s, and the idea of intelligent machines dating back to ancient Greece. What seems like a dramatic leap today will likely be seen as a small step in the long term. Zeno's paradox reminds us that dividing complex goals into manageable tasks drives progress, and that overthinking can delay action. Similarly, AI innovation involves exploration, creative problem-solving, and often taking unconventional paths. It's not just about advancing technology but fostering social innovation—finding new ways to collaborate, include diverse voices, and regulate effectively. Each step, no matter how small, contributes to building more sophisticated and inclusive systems.

Paradoxes also show how our actions can lead to unexpected outcomes. For example, improving car fuel efficiency may seem environmentally friendly, but it can encourage more driving, which reduces the overall benefit. Similarly, focusing too much on making AI systems highly accurate can unintentionally result in systems that are harder to understand, less reliable, or more biased. This overemphasis on accuracy can lead to unintended consequences, such as AI systems that prioritize technical precision over fairness, transparency, or practical usefulness. To address this, we need a balanced approach that considers not just how accurate systems are, but how they affect people and society as a whole.

Key Takeaways and Reflections

Progress in AI is inseparable from human creativity and intellect. Every step forward in AI development relies on our ideas, imagination, and decisions. As AI evolves, it redefines tasks that require uniquely human traits such as emotional intelligence, ethical reasoning, and creativity. The fundamental AI paradox, that the more AI achieves, the more it underscores the unique and irreplaceable nature of human intelligence, reminds us of our crucial role in shaping AI's impact. This calls for active engagement and adaptation as we prepare for the transformations AI will bring.

As Pablo Picasso once said, "Computers are useless. They can only give you answers." While AI provides answers to complex problems, it also pushes us to ask sharper, more meaningful questions. The AI paradox encourages us to anticipate unexpected outcomes and approach challenges from multiple perspectives. These reflections allow us to make thoughtful and responsible decisions that guide us toward a future where AI supports and enhances human values.

In summary, the fundamental AI paradox underscores the unique qualities of human intelligence that remain beyond AI's reach. While AI transforms the world around us, it relies on human creativity and ethical judgment to advance. To thrive in this evolving landscape, we must embrace critical thinking, adapt to new challenges, and consider the broader impacts of AI. By understanding and engaging with paradoxes like these, we can shape a future where AI aligns with our collective goals and aspirations.

2

The Agreement Paradox

DEFINING AI

A COUPLE of years ago, during a coffee break in one of the many high-level AI policy expert groups I participate in, I struck up a conversation with a fellow expert. It was the first meeting of this group, and we were all eager to contribute to establishing principles and guidelines for the trustworthy development and use of AI. My colleague started the conversation with an enthusiastic remark: "I'm really happy to be here; this is such an important topic. We used to have software, but now we have AI." I nearly choked on my coffee! I mumbled, "You know, AI is software ..." to which, after a visible moment of mental processing, they responded, "Yes, but now we can open up the software and put AI inside!" I often share this story when discussing the varied understandings of AI and how policy is addressing it. At that moment, I was flabbergasted and concerned: How could a member of a group tasked with defining AI guidelines hold such a view of the technology we were supposed to govern? Could such a person contribute anything meaningful to the discussion? My stance has since evolved. Developing AI governance and policy is not just about

assembling the most knowledgeable AI experts to draft regulations. It is about the insights that emerge from the interaction
between those who understand the technical development of
these systems and those who experience the real-world consequences of their use. It is about bridging cutting-edge scientific
research with the hopes, fears, and daily realities of the *people
on the street*. Understanding AI requires embracing the many
different conceptions of it: from the computational "purists" to
those who see it as a "magic" ingredient you can add to software,
from the developers to the users, and from the enthusiasts to
the skeptics.

Even though everyone is talking about AI, if you closely
follow developments in this field, you quickly realize that AI
means different things to different people. So, what exactly is
this "AI" that everyone is discussing? What are we really concerned about? What do we want to regulate? Is AI a technology
or a field of science? Is it a tool or an entity? Is AI out of control,
or are we losing control over those who control it? Is governance and regulation of AI a necessity or a futile effort? Will
AI extinguish humankind, or will it solve all our problems? So
many questions, so many opinions. At the same time, many
products and services, from washing machines to cars and personal assistants, are now branded as *AI-powered*.[1] This trend
reflects both genuine advancements in AI and strategic marketing efforts to capitalize on the AI hype.[2] It seems that all of a
sudden, every digital system is AI, AI is used everywhere and
can solve anything, and the more we will use AI in any field, the
better.

Opinions about what AI is seem to fall into two broad
extremes. On one side, AI is seen as a near-magical solution that
will solve all problems, even if we don't fully understand how
it works—an almost omnipotent force poised to revolutionize

or even control the world. On the other extreme, it's viewed as "business as usual," a natural next step in the ongoing evolution of technology, where focusing on the capabilities of AI to enhance efficiency and automate tasks will propel digitalization from mere information management to intelligent interaction with the world.

However, these two perspectives don't capture the full range of views on AI. Many opinions fall somewhere in between, recognizing both the potential and the limitations of AI. Some focus on AI's ethical and societal implications, while others are concerned with its technical challenges or economic impact. Table 2.1 gives an overview of the many different ways AI is being defined by policymakers, authors, and media outlets.

As authors Arvind Narayanan and Sayash Kapoor argue in *AI Snake Oil*,[3] the hype surrounding AI is often fueled by exaggerated claims from companies, nonreproducible studies from researchers, and sensationalized media coverage. This leads to a distorted view of AI's capabilities, fostering misconceptions. Narayanan and Kapoor emphasize the need for a more critical and informed discourse to better understand AI's true potential and limitations, advocating for education to counter the hype and enable more responsible decisions about AI's role in society. This misconception contributes to a growing lack of understanding about what we truly mean when we talk about AI, which is the gist of our next paradox:

The Agreement Paradox
The more we explore AI, the harder it becomes to agree on its definition.

As has happened before with the term *digitalization*, AI is becoming an empty signifier,[4] a term that suggests a specific

TABLE 2.1. How AI definitions focus on different issues depending on the source.

Source	Definition	Focus issues
European Union[a]	A machine-based system that is designed to operate with varying levels of autonomy and that may exhibit adaptiveness after deployment, and that, for explicit or implicit objectives, infers, from the input it receives, how to generate outputs such as predictions, content, recommendations, or decisions that can influence physical or virtual environments.	Autonomy, adaptiveness, decision-making
Stanford Center for Human AI (HAI)[b]	The science and engineering of making computers behave in ways that, until recently, we thought required human intelligence.	Replication, engineering, learning
Organization for Economic Cooperation and Development (OECD)[c]	A transformative technology capable of tasks that typically require humanlike intelligence, such as understanding language, recognizing patterns, and problem-solving.	Problem-solving, humanlike tasks
U.S. Department of State[d]	A machine-based system that can, for a given set of human-defined objectives, make predictions, recommendations, or decisions influencing real or virtual environments.	Prediction, recommendation, decision-making
International Standards Organization (ISO)[e]	A technical and scientific field devoted to the engineered system that generates outputs such as content, forecasts, recommendations.	Engineering, forecasts

Continued on next page

TABLE 2.1. (*continued*)

Source	Definition	Focus issues
Artificial Intelligence: A Modern Approach by Stuart Russell and Peter Norvig[f]	The study of agents that receive percepts from the environment and perform actions.	Interaction, action
The Master Algorithm by Pedro Domingos[g]	The ultimate learning machine capable of deriving all knowledge from data.	Data-driven learning
Artificial Intelligence: A Guide for Thinking Humans by Melanie Mitchell[h]	Machines that do things that seem intelligent.	Intelligent appearance, task-specific
The Guardian[i]	The simulation of human intelligence in machines that are programmed to think like humans and mimic their actions.	Simulation, mimicking
ABC News[j]	The simulation of the human capacity to think and learn for the sake of performing tasks.	Simulation, task performance

[a] https://artificialintelligenceact.eu/article/3
[b] https://hai.stanford.edu/sites/default/files/2023-03/AI-Key-Terms-Glossary-Definition.pdf
[c] https://www.oecd.org/en/topics/policy-issues/artificial-intelligence.html
[d] US Department of State (archived content 2021-2025): https://2021-2025.state.gov/artificial-intelligence/.
[e] https://www.iso.org/artificial-intelligence
[f] Stuart Russell and Peter Norvig, *Artificial Intelligence: A Modern Approach*, 4th ed., Pearson 2020.
[g] Pedro Domingos, *The Master Algorithm*, Basic Books 2015.
[h] Melanie Mitchell, *Artificial Intelligence: A Guide for Thinking Humans*, Farrar, Straus & Giroux 2019.
[i] https://www.theguardian.com/technology/2023/feb/24/ai-artificial-intelligence-chatbots-to-deepfakes
[j] https://abcnews.go.com/Technology/what-is-artificial-intelligence-ai/story?id=99919927

meaning but resists a clear definition, thereby maximizing its suggestive power. In the same way, everybody is referring to "AI," assuming that others (media, policymakers, and the general public) understand what it means. The problem of AI becoming an empty signifier goes beyond mere miscommunication; it can also lead to credibility issues because conflicting interpretations erode trust in AI initiatives, in particular when expectations are not met.

This vagueness is not just confusing; it's a deliberate tool of power. As we'll explore further in chapter 6, the ambiguity around AI is intentionally maintained by those who seek to control its narrative. By keeping the definition unclear, they not only shape the meaning to suit their interests but also create the illusion that the current trajectory of AI is inevitable. This fosters a sense of magic and powerlessness in others, making it seem as though the rise of AI is beyond challenge or change. This strategic ambiguity ensures that those in power maintain control, reinforcing their dominance while others feel both uninformed and helpless.

The fact that AI is a vague concept with different meanings for different people presents significant challenges for policymaking. A lack of clarity hampers the formation of an informed understanding of AI and its societal and ethical implications. If we don't have a clear definition of what we are trying to govern, how can we be confident that our policies will be effective? And if we cannot be sure, why even attempt regulation? Disparate understandings of AI create friction between stakeholders, with technologists advocating for innovation and policymakers pushing for stricter regulations, leading to policies that fail to address critical risks while also fostering unnecessary fear and resistance to AI adoption.

What Is AI?

Things used to be simple(r). AI was once a relatively obscure field within computer science, with occasional incursions into philosophy or cognition. Although researchers were exploring and defining AI in various ways, these discussions largely remained within academic circles. However, even then, the real-world impact of AI was clear when AI was put to practical use. I first encountered this impact in 1986, when I developed my very first AI application. Commissioned by the now-defunct Ministry of Public Works in Portugal, the project aimed to create a system for the fair and uniform allocation of social housing. In line with the technology of the time, I built an expert system: a rule-based implementation of the norms, laws, and best practices followed by civil servants responsible for these decisions. Fresh out of university, I developed the system on my own, feeling a mix of pride in their trust in my abilities and concern about the potential errors I might make. The system was designed to take a list of available housing units and a list of clients in need, and then apply formal rules to prioritize the allocation of houses. Everything seemed to be working well until one of the first real-world tests, when the system assigned a house that was actually already occupied by someone else. This was a swift and harsh lesson for me: even if the system is technically correct, it is only as reliable as the data it uses, data that, in this case, was not up to date.

Fast forward several decades to the present day, and the social and ethical impact of AI development and use is a fundamental part of the work in the field, with many organizations and international agencies tasked with defining and implementing principles and policies. This is also the area of my own research and policy work. As AI has become more integrated into various aspects of life, its implications have grown more complex and far-reaching. Despite all the advancements, a crucial question remains: how much can we trust the accuracy of AI system decisions, and to what extent do they truly reflect reality? This issue has far-reaching consequences, not only for the data these systems rely on but also for the accountability of those who deploy them. The question of what AI actually is has become more pressing, and the answer is not as straightforward as it once seemed.

To understand the role of AI in society and its contribution to the broader digital transformation, we first need to grasp what AI truly is and what, if anything, sets it apart from other technologies. Back in the 1950s, the term artificial intelligence was initially introduced by John McCarthy and colleagues in a proposal for a summer conference to address *"the conjecture that every aspect of learning or any other feature of intelligence can in principle be so precisely described that a machine can be made to simulate it."* The recent success and hype surrounding AI have led to a wide range of definitions. For example, a very simplified definition from an early European Commission white paper[5] describes AI as *"a collection of technologies that combine data, algorithms, and computing power,"* which could arguably apply to almost any computer program. On the other end of the definitional spectrum, Elon Musk has referred to AI as "godlike,"[6] and historian Yuval Noah Harari has suggested that AI could even give rise to a new religion.[7]

As we navigate this landscape, it is important to establish some level of common understanding. From the initial definition of AI as an artificial system that can emulate human intelligence, today AI is seen in many different ways: as an artificial information processing system, as an augmentation of human capabilities, or even as a potential form of superior intelligence.

Establishing a common understanding of AI is crucial for addressing its impact on society and ensuring that policies and discussions are based on a shared definition of what AI actually is. Without a clear understanding of what AI can and cannot do, efforts to govern, regulate, and responsibly develop AI risk becoming ineffective. So, let's clarify the concept. Originally, as McCarthy defined it, AI was a field of research focused on understanding human intelligence by building computational systems. The driving force was not the technology itself but the pursuit of insight into human cognition. As the field progressed and began developing systems that could mimic aspects of reasoning, planning, natural language processing, and learning, new computational techniques emerged to specialize in these functions. Techniques such as logic programming, neural networks, and search algorithms became valuable not just for theoretical exploration but for practical applications. Among these, neural networks proved to be the most scalable and also particularly suited for learning tasks, which is why they have become the most prominent AI methods today. In the 1980s, neural networks may have had a few thousand inputs, a handful of neuron layers, and a couple of outputs. In contrast, modern large language models (LLMs) consist of billions of input parameters, thousands of layers, and outputs that are equally vast. The immense complexity of these networks makes it nearly impossible to fully understand how a specific output is generated. This opacity contributes to the perception of AI as "magic," where the inner workings are not easily understood even by the experts.

This complexity and lack of transparency create challenges in how we view and manage AI. Depending on the focus and context, various perspectives on AI's societal impact emerge. These perspectives range from concerns about ethical responsibility and bias in AI systems to the implications of AI for jobs, privacy, and decision-making autonomy. Understanding what AI can and cannot do is critical for ensuring that these conversations remain grounded in reality and for developing policies that effectively address AI's broader influence on society, and therefore it is important to differentiate between different understandings and uses of AI:[8]

- A (computational) technology that is able to infer patterns and possibly draw conclusions from data based on algorithms. This view is sometimes referred to as artificial information processing and often brings under the general denomination of AI many different technologies, from robotics to the Internet of Things, and from data analytics to cybersecurity, the result of which is that everything is considered to be AI.
- A field of scientific research focusing on understanding and developing intelligent behavior in machines. This (mostly academic) perspective involves studying theories and methods that enable machines to adapt, interact, and operate autonomously. It emphasizes the foundational principles behind machine learning, perception, and decision-making, aiming to advance the theoretical knowledge that drives AI development.
- An autonomous entity or agent, a portrayal common in media and science fiction, where AI is seen as a self-governing force with near-magical, all-knowing, and all-powerful qualities. This perspective often suggests that AI operates beyond human control, fueling dystopian

fears that it acts on us rather than being something we can manage or influence.

The key point here is not to select one "correct" definition but rather to understand that AI is defined by a series of complex, shifting perspectives. AI resists a rigid definition because it is a dynamic concept, shaped by ongoing technological advancements and changing societal expectations. As a socio-technical system, AI is deeply interconnected with human decision-making, ethics, and governance. Nevertheless, rather than accepting that the concept of AI is an empty signifier, we remain aware of the forces that shape specific interpretations of AI, particularly those promoting a "magical" or uncontrollable view. The core message of this book, to which we will return several times, is that AI is designed: it is created and defined by people. In the following sections, we will further explore the multifaceted nature of AI.

As we explore the complexities of AI in this book, it becomes clear that a singular, precise definition of AI remains elusive. These varying perspectives reflect the evolving nature of AI and the different contexts in which it is applied. In the following subsections, I will describe different frames used to understand how AI interacts with the world and society. These frames include seeing AI as *information processing*, based on its capabilities to analyze and handle data; as a *decision-making* system, by which it makes choices based on input; and as an *interactive* system, given its capabilities to shape and be shaped by humans, society, and the environment. This approach makes it easier to understand the various ways AI is conceived and used, and sets the stage for exploring its broader impact as it becomes increasingly integrated into our daily lives.

AI Is _Artificial Information_ Processing

AI can be understood as an advanced form of information processing, encompassing tasks like collecting, analyzing, and recognizing patterns in data. Like other information processing software, AI relies on certain generic qualities to function effectively. However, these qualities take on heightened importance in AI because of its role in making decisions, predictions, and recommendations that directly affect real-world and virtual environments. For instance, _robustness_, the ability to handle errors and inconsistencies in data, is critical in AI because real-world data is often messy or unpredictable. Without robustness, AI systems would fail in practical applications. Similarly, _adaptability_ becomes vital in AI as it allows systems to learn and adjust to new environments or data, expanding their utility across diverse contexts. _Perception_ and _interaction_ capabilities, such as interpreting speech or images, are essential in AI to enable meaningful responses to users and environments. Cognitive abilities, like _reasoning_ and _decision-making_, are foundational in AI for tackling complex problems that require logical and contextual understanding, far beyond the static rules of traditional software. Lastly, _scalability_ and _efficiency_, critical for all software, are especially crucial in AI to handle massive datasets and computational complexity without prohibitive resource demands. In AI, these qualities are not just technical necessities; they enable the system to fulfill its promise of flexibility, learning, and responsiveness, which sets it apart from conventional software solutions.

Within this large field, AI technologies can be categorized into the following types, each reflecting distinct approaches to how systems process information and solve problems:

- **Symbolic, or Rule-Based, AI**: This type of AI uses explicitly defined rules and logical reasoning to make decisions or solve problems, rather than learning from data. Often referred to as "Good Old-Fashioned AI" (GOFAI), it is one of the earliest approaches to AI but remains highly relevant in practical applications today, such as tax-preparation systems. These systems rely on predefined *if-then* rules derived from tax laws to assist users in filing their taxes. For example, they might use logic like *"if income is below a certain threshold, apply this tax credit,"* ensuring both accuracy and compliance with regulations. This approach is particularly effective in areas like taxes, where clear rules and transparency are crucial for helping users understand the reasoning behind decisions.
- **Machine Learning, Including Deep Learning**: Machine learning (ML) is a subset of AI that involves algorithms learning from data to make decisions or predictions. Deep learning, a subset of machine learning, uses neural networks with many layers (hence "deep") to analyze complex patterns in large amounts of data. Technically, this type of AI learns its rules from statistical correlations based on the data, rather than hand-crafted rules based on expert knowledge. This category has been responsible for many of the recent advancements in AI, including image and speech recognition.
- **Generative AI and Foundational Models**: Generative AI (GenAI) is a subset of machine learning that describes any type of artificial intelligence (AI) capable of producing new text, images, video, or audio. Although it is a type of ML, its primary focus is on generating unique, creative outputs rather than performing analytical tasks

typical of other ML techniques. Foundational models, often discussed in the context of generative AI, are large-scale models that can be adapted to a wide range of tasks and datasets. Models like GPT (generative pretrained transformer) and DALL-E have demonstrated impressive capabilities in generating humanlike text and realistic images, respectively. These models serve as foundational platforms for building more specialized AI applications across various domains.

There are many more theories and techniques within the vast field of AI. For a nontechnical introduction to AI, I strongly recommended the book *Artificial Intelligence: A Guide for Thinking Humans* by computer scientist Melanie Mitchell.[9] Also, the main types of systems above should not be understood as orthogonal to each other, but more as points in a nexus. As Professor Paul Lukowicz, a machine learning expert from DFKI, the German AI Research Institute, says: *"There is no sharp difference between 'classic' machine learning and general purpose models; it is a continuum."* That is, generating a prediction, a piece of text, or even a figure involves fundamentally similar processes. These outputs, while differing in form, all stem from the same core principles of learning from data and applying that knowledge to create meaningful results. Figure 2.1 illustrates the continuum of AI development, highlighting their progression from rule-based systems to generative and general-purpose models. Basically, all AI models take in data and produce outputs. What distinguishes the different approaches is how they process that input data, how the models are built, and how broadly the results of those models can be applied. Earlier AI approaches, such as rule-based systems and classic machine learning, depend on curated, task-specific data

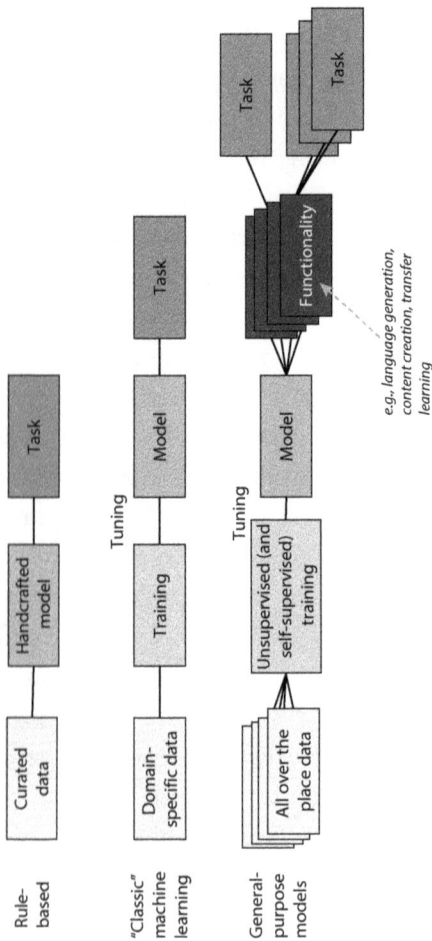

Rule-based

Curated data — Handcrafted model — Task

"Classic" machine learning

Domain-specific data — Training — Tuning — Model — Task

General-purpose models

All over the place data — Unsupervised (and self-supervised) training — Tuning — Model — Functionality — Task / Task

e.g., language generation, content creation, transfer learning

FIGURE 2.1. Different types of AI system

and handcrafted models designed for particular applications. These models are typically constrained to narrow domains, where they excel due to their specialized design. In contrast, newer methods like generative AI and foundational models utilize vast and diverse datasets, allowing them to process a wide variety of information. These models are built to adapt across multiple tasks, evolving through advanced training techniques and fine-tuning, which enable them to generate new outputs or make decisions in many different contexts.

Currently, much work is being done to combine the benefits of these different approaches. One promising avenue is **neurosymbolic AI**, which merges the pattern recognition capabilities of neural networks with the logical reasoning of symbolic AI. This combination allows systems to both learn from data and make explainable, rule-based decisions. For instance, in medical diagnosis, a neural network might identify patterns in X-ray images that suggest a potential abnormality. A symbolic reasoning layer can then apply established medical knowledge and rules, such as *"If a mass is detected in the lungs and the patient has a history of smoking, infer a higher risk of lung cancer."* This layered approach not only improves accuracy but also provides an explanation of the decision-making process, making it easier for doctors to trust and verify the results. Neurosymbolic AI also has applications in fields like legal analysis, where neural networks extract relevant case details and symbolic reasoning applies legal principles to provide transparent recommendations. By combining learning with reasoning, neurosymbolic AI bridges the gap between data-driven insights and humanlike interpretability, opening doors to more robust and trustworthy AI systems.

Many other scientific directions and techniques are a part of what is commonly understood as AI, including reasoning (for example, planning, scheduling, knowledge representation, search, and optimization) or robotics (for example, control, perception, sensors and actuators, and the integration of physical/hardware components). Overemphasizing the distinctions between various AI approaches can lead to fragmented research, a lack of integrated understanding, ineffective policymaking, and a lack of appreciation for how these different methods can complement each other in practical applications, potentially limiting the effectiveness and adaptability of AI solutions.

For instance, combining reasoning with robotics can create adaptive robots that navigate environments and plan actions intelligently. Similarly, merging neural networks with optimization techniques produces systems that recognize patterns and make efficient decisions. Collaboration and integration across AI methods are key to unlocking its full potential. By recognizing how these approaches complement each other, researchers and policymakers can develop more effective and adaptable solutions to a wide range of challenges.

AI Is _Augmented_ (social) _Interaction_

Considering the impact of algorithms, systems, and applications, AI is much more than information processing alone. Nevertheless, it is not "magic" but it is also not "business as usual." These narratives are dangerous because they portray the idea that there is nothing we can do to change the effects. The "business as usual" narrative can lead to inaction and a sense that all will stay as it is anyway. The "magic" narrative, which is fed by science fiction and by the popular press, gives the impression that nothing can be done against such an all-knowing entity that rules over us in possibly unexpected ways and that will either solve all our problems or destroy the world in the process. But if AI is neither magic nor business as usual, how best can we describe AI in order to take into account not only its capabilities but also its societal implications?

Yes, AI is based on algorithms, but then so is any computer program and most of the technologies around us. Nevertheless, the concept of "algorithm" is achieving magical proportions, used right and left to signify many things, de facto seen as a synonym to AI. The easiest way to understand an algorithm is as a recipe, a set of precise rules to achieve a certain result. Every time you multiply two numbers, you are using an algorithm, as much as you are when you are baking an apple pie. However, by itself, the recipe has never turned into an apple pie; and the end result of your pie has as much to do with your baking skills and your choice of ingredients as with the choice of a specific recipe. The same applies to AI algorithms: the behavior and results of the system largely depend on its input data, and on the choices made by those who developed, trained, and selected the algorithm. In the same way as we have the choice to use organic apples to make our pie, in AI we also must consider our choices of which models and data to use, who to include in the design and considerations about impact, and how these choices respect and ensure fairness, privacy, transparency, and all other values we hold dear.

AI is first and foremost technology that can automate (simple, lesser) tasks and decision-making processes. At the present, AI systems are largely incapable of understanding meaning and the context of their operation and results. At the same time, considering its societal impact and need for human contribution, AI is much more than an automation technique. When considering effects and the governance thereof, the technology, or the artifact that embeds that technology, cannot be separated from the socio-technical ecosystem of which it is a component. In this sense, the first important distinction to be made lies in understanding AI as either a computer system or a socio-technical entity: as a computer system, AI usually refers to computer systems that exhibit autonomous, adaptable, and interactive properties and are capable of decision-making based on environmental data to achieve complex goals. As a socio-technical system, AI extends beyond just software or hardware, encompassing its interactions and influences within societal and technical frameworks.

Understanding and defining AI as a socio-technical ecosystem enables us to recognize the different roles, the interaction between people and technology, and how such complex infrastructures affect and are affected by society and by human behavior.[10] For example, the Dutch child benefits scandal illustrates that AI-driven decision-making is not just about technology; it is about the intersection of automated systems with social and institutional structures. In this case, the Dutch Tax and Customs Administration employed an algorithmic system to detect fraudulent childcare benefit claims based on their own criteria and risk variables, without allowing citizen input or redress. As a result, families with dual nationalities and lower incomes were disproportionately targeted, leading to wrongful accusations and severe financial and emotional distress for thousands.[11] The scandal ultimately led to the resignation of the Dutch government in 2021.[12] Automating decisions without social safeguards can increase inequality and reduce public trust, showing the need for human-centered AI. AI is not just about automating actions or learning from data; it is shaped by who controls it, what knowledge it uses, and how those affected can participate in the process.

AI Is _Ambiguous Interpretation_

The growing hype around AI is blurring its definition and shoving into the same heap concepts and applications of many different sorts. Ensuring the responsible development and use of AI is becoming a main direction in AI research and practice. Governments, corporations, and international organizations alike are coming forward with proposals and declarations of their commitment to an accountable, responsible, transparent approach to AI, where human values and ethical principles lead. Currently, more than six hundred AI-related policy recommendations, guidelines, or strategy reports have been released by prominent intergovernmental organizations, professional bodies, national-level committees and other public organizations, nongovernmental organizations, and private for-profit companies. Even though there are many differences, a study of the global landscape of AI ethics guidelines shows a global convergence around five ethical principles: transparency, justice and fairness, non-maleficence, responsibility, and privacy.[13] Nevertheless, this proliferation of guidelines and principles is not really conducive to a shared understanding and collective action. A much-needed first step in the responsible development and use of AI is to ensure a proper AI narrative, one that demystifies its capabilities, minimizes both overselling and underselling of AI-driven solutions, and enables wide and inclusive participation in the discussion on the role of AI in society. Understanding the capabilities and addressing the risks of AI requires that all of us—from developers to policymakers, from providers to end users and bystanders—have a clear understanding of what AI is, how it is applied, and what the opportunities and risks involved are.

Current AI techniques are largely grounded in data-driven strategies, especially in the realm of machine learning, generative AI, and foundation models, which use statistical methods to empower computers to recognize aspects of their environment, excelling in interpreting images or text datasets. These advanced models excel at pattern recognition, learning from extensive datasets (often millions of examples) to discern patterns and make predictions about new, unseen data. Despite their impressive capabilities, these approaches don't fully embody what we would consider "true intelligence." The generative AI and foundation models, though transformative, are plagued by challenges such as mistakes (also called hallucinations), bias, lack of transparency, and issues with explainability and inclusivity. The complexity of these algorithms often makes them impenetrable for direct human understanding, and their reliance on historical data risks entrenching, replicating, and enlarging past biases and prejudices. Indeed, data is always about the past, and decisions on which, how, when, and why to collect and maintain data fundamentally influence the availability and quality of data. Those who have the power to decide on data have the power to determine how AI systems will be designed, deployed, and used.

Much has been said about the dangers of biased data and discriminating applications. Minimizing or eliminating discriminatory bias or unfair outcomes is more than excluding the use of low-quality data. The design of any artifact, such as an AI system, is in itself an accumulation of choices, and choices are biased by nature because they involve selecting one option over another. Most important, it starts with the current reliance on data as a measure of what can be done. Increasingly, we are seeing that the availability of data (or the possibility to access data) is taken as a guiding criterion to solving societal issues. If there are data, it is a problem we can address, but if there are no data, there is no problem. This is intrinsically related to power and to power structures. Those who own or control the data can decide which problems are worth addressing, and therefore can shape not only how AI is being developed and used but also which technologies to use and what values to prioritize. Those in power over data are shaping the way we live with AI and how our future societies will look.

Nevertheless, attention for the societal, environmental, and climate costs of AI systems is increasing. All these must be included in any effort to ensure the responsible development and use of AI. A responsible, ethical approach to AI will ensure transparency about how data is used, adaptation is done, responsibility for the level of automation on which the system is able to reason, and accountability for the results and the principles that guide its interactions with others, most importantly with people. In addition, and above all, a responsible approach to AI makes clear that AI systems are artifacts manufactured by people for some purpose, and that those who make them have the power to decide on the use of AI. It is time to discuss how power structures determine AI and how AI establishes and maintains power structures, and on the balance between those who benefit from and those who are harmed by the use of AI.[14]

AI Is Agent Interaction

Conceptualizing AI as agent interaction focuses on how AI systems, whether as individual agent or as part of a multiagent framework—engage with one another and with humans to accomplish tasks, solve problems, and make decisions.

The early boom of multiagent systems (MAS), which lasted from the mid-1990s to around 2010–2015, produced significant scientific results but struggled to deliver widespread real-world applications. Some notable successes included the use of game theory in security games for airport defense, automated high-frequency trading systems, and agent-based modeling to simulate socio-technical systems during the 2008 financial crisis and the 2020 COVID-19 pandemic.[15] The original vision of multiagent systems, where autonomous agents interact and collaborate, remains largely unrealized, even as reactive single agents like Alexa and Siri have become common. With the advent of LLMs, the concept of agentic AI, focusing on integrating generative AI and LLMs into autonomous systems to enhance creativity, decision-making, and interaction, has generated significant buzz, though much of the hype appears to be driven by marketing.

The growing fascination with agentic AI reflects high expectations for AI systems to act as capable assistants, handling specific human tasks such as scheduling meetings, booking travel, drafting documents, or generating creative content. These agents are envisioned to streamline workflows, support decision-making, and thus reduce the need for human intervention. However, this ideal of fully autonomous agents often exaggerates their capabilities, suggesting they can navigate complex decisions without human oversight. Although such a vision is appealing, it places too much value on AI capabilities of acting autonomously while overlooking the critical importance of coordination and collaboration. In fact, AI agents should be understood less in the light of philosophical agency and more as proxies, that is, task-specific tools designed for clear objectives. As such, using overly complex systems like LLMs for narrowly defined tasks is inefficient. Simpler, specialized models can achieve the same outcomes with fewer resources. For example, just as we would not expect to discuss philosophy with a travel operator, a travel-planning AI should focus solely on bookings and itineraries without the computational overhead of broader generalization or unrelated tasks.

AI agents face many other practical limitations, such as struggling with commonsense reasoning and generalization, which makes them unsuitable for tasks requiring nuanced decision-making or ethical judgment. Most important, the focus on agent autonomy misses the true challenge of AI development. Although simple systems like thermostats are fully autonomous, they do not represent meaningful AI innovation. Instead of focusing solely on isolated autonomy, a more nuanced perspective recognizes the need for agents to function effectively within interconnected systems, working alongside humans and other agents to achieve shared goals.

The real contribution of the agent paradigm in AI may not be for creating AI assistants, but as a new way to design systems. Instead of building one large, all-powerful model, agents can work as modular parts of a larger system, each handling a specific task. These agents collaborate, share information, and adapt to changes, creating systems that are flexible, efficient, and easier to manage. For example, in a transportation system, one agent could optimize traffic flow, another plan routes, and a third report incidents in real time. Together, they form a system that handles complex challenges more effectively than a monolithic model. This approach allows developers to update or improve individual parts without disrupting the whole system. It also makes AI systems easier to understand and audit, increasing transparency and accountability. By focusing on collaboration and specialization, the agent paradigm offers a practical and innovative way to design AI systems that are better suited for real-world problems. This approach not only offers a more sustainable alternative to current general-purpose

AI models but also aligns better with the complexities of real-world applications. By focusing on specialization and collaboration, compositional AI creates systems that are flexible, efficient, and more in tune with human values.

What Do We Want AI to Be?

In whatever way we define AI, it is crucial to understand that it is an artifact, that is, something created by people. Thus, the essential question is not what AI is, but rather **what we want AI to be**. As I have argued many times,[16] since we build it, we control and are responsible for its trajectory and choices.

This question reshapes our relationship with AI. Currently, there is effectively a select group that asks, decides, and builds what AI should be. This is a small, nonrepresentative, and noninclusive group that essentially dictates AI's nature and role for everyone. Their decisions and actions de facto determine AI's trajectory, influencing how it integrates into and affects the broader society. As we will see in chapter 5, there are vested interests in keeping the meaning of AI vague and thus the chances of successful regulation slim.

Reflecting on what we want AI to be forces us to confront a complex array of questions that shape its development and role in society. Should AI mimic humans and, if so, are we prepared to address the ethical and philosophical implications of choosing which human traits are ideal? But what does "being like a human" even mean in the context of a machine? Which human are we considering, or which aspects of humanity are worth replicating? Is it about voice, appearance, behavior?[17]

As AI advances, there's an increasing focus on making machines more humanlike driven by the desire to create more intuitive and effective interactions between humans and machines. For instance, AI that can understand and replicate human speech patterns more accurately can be useful in customer service, therapy, or accessibility tools for those with disabilities. There are, thus, good reasons for enhancing these aspects of AI, primarily to improve accessibility, provide better user experiences, and expand the range of applications where AI can be helpful. For example, when developed and deployed responsibly, a voice that sounds natural and empathetic can make interactions with AI feel more personal and engaging in fields like healthcare, where it might be used in therapy, education, or even companionship for individuals who are isolated or have specific needs. However, this push toward humanlike AI also raises significant ethical concerns, particularly around the potential for impersonation. Regulations that forbid AI from impersonating humans seek to prevent deception and protect individuals from being misled by machines that could be used to imitate real people. Such regulations are crucial as AI's ability to mimic human voices and behaviors becomes more sophisticated. Without these safeguards, the risk of AI being used for malicious purposes—such as fraud, identity theft, or manipulating public opinion—grows significantly. A particularly alarming development in this context is the rise of deepfakes, AI-generated content that can convincingly mimic the appearance and voice of real people. Deepfakes can be used for harmful activities, such as spreading misinformation, creating fake news, or damaging someone's reputation. This underscores the urgent need for clear regulations and ethical guidelines to ensure that AI technology is used responsibly and does not undermine trust in digital content.

Another critical choice to consider, alongside or as an alternative to the question of humanlike AI, is whether we want to view and use AI as a tool. This perspective shifts the focus from trying to replicate human traits to leveraging AI for specific, functional purposes. Viewing AI as a tool emphasizes its practical applications, but it also brings with it important questions and implications.[18] Tools, by their very nature, come with inherent affordances and limitations; they can enable certain actions while restricting others. For instance, even simple tools like scissors are designed for right-handed users, making them difficult or even unusable for left-handed people like me. That is, the design of a tool can, unintentionally, exclude certain groups, leading to unequal access and usability. Similarly, in the context of AI, the decisions about how these tools are designed and who has access to them can create or reinforce existing inequalities. Who gets to decide how AI tools are used, and who has access to them? These decisions are not just technical; they are deeply intertwined with social and ethical considerations. The power dynamics involved in these choices could determine who benefits from AI and who might be left behind. For instance, if AI tools are developed primarily by and for certain groups, they will reinforce existing inequalities, giving more power and advantages to those who already hold them while marginalizing others. Conversely, if AI is designed and governed with inclusivity in mind, it has the potential to empower a broader range of people and communities. Considering AI as a tool rather than as a humanlike entity highlights the importance of transparency and accountability in its use. Unlike humanlike AI, which might raise concerns about deception and trust, AI as a tool can be designed to be more transparent in its operations and decisions, allowing users to understand its capabilities and limitations more clearly.

Another crucial consideration is whether we want AI to simulate the world or to affect it.[19] If AI is designed to simulate the world, whether modeling complex systems, predicting outcomes, or offering insights, it is essential to question the accuracy and the implications of these simulations. Who ensures that the data used is representative and unbiased? Who validates the insights provided by these simulations, and what checks are in place to prevent them from being misused? Simulations can offer powerful tools for understanding and decision-making, but they are only as reliable as the data and algorithms behind them. Here again, transparency and accountability are critical: users need to know how these simulations are constructed, and developers must be held accountable for the outcomes they produce. On the other hand, if AI is designed to act upon the world, making decisions that directly affect people's lives, such as in healthcare, criminal justice, or autonomous driving, the stakes are even higher. The question of accountability becomes much more complex. When AI systems make decisions autonomously, who is responsible for those decisions? Is it the developers who created the algorithms, the organizations that deploy the AI, the end users who rely on the system's outputs, or the policymakers who allow such situations to happen? This murkiness in responsibility can lead to significant ethical and legal challenges, particularly when AI decisions result in harm or injustice. Furthermore, the use of AI in decision-making processes brings up concerns about the erosion of human judgment. As we increasingly rely on AI to make complex decisions, there is a risk that human oversight might diminish, leading to decisions that are technically sound but lack the nuance and empathy that human judgment provides. This is particularly concerning in areas where ethical considerations are paramount, such as in medical diagnoses or sentencing in the criminal justice system.

The scope and impact of AI development raises significant concerns in an era where progress relies heavily on massive datasets and computational power. These efforts come with high costs, including resource consumption (electricity, water, rare metals) and human labor. For example, the International Energy Agency projects that by 2026, data centers will consume as much electricity as Japan. Additionally, water use in tech companies' data centers is rapidly increasing, with Google's usage up 20 percent and Microsoft's up 34 percent in 2022 alone. These numbers, however, don't provide the full picture. Data on AI's specific resource use is limited, making it difficult to separate its impact from other data center activities, such as cryptocurrency mining. Although AI can help save resources in some areas, it may also lead to yet another paradox, the Jevons paradox, which means that making something more efficient can actually cause people to use it more, increasing overall consumption. For example, widening a freeway might reduce traffic congestion and save fuel at first, but the easier driving conditions will attract more cars, eventually leading to higher fuel use. Similarly, as AI becomes more efficient, it could encourage more widespread use, increasing resource demands overall. Monolithic, centralized AI models exacerbate these issues. They are not only resource-heavy and difficult to verify but also often apply a "one size fits all" approach, benefiting those in control while neglecting local needs. Moreover, the human cost of AI includes poorly compensated labor for tasks like data labeling, which perpetuates global inequalities. Automation threatens these jobs, further widening the digital divide. A shift to modular, compositional AI systems can offer a more sustainable path. These approaches rely on smaller datasets, independent development, and layered architectures. By limiting resource consumption and reducing reliance on large-scale human labor, they enhance efficiency, adaptability, and fairness while addressing the ethical and environmental challenges of traditional AI models.

Returning to the fundamental questions in AI development: Should AI mimic our actions or follow our instructions? Do as we do, or do as we say? If AI is designed to replicate human behaviors and decisions, there is a risk of perpetuating our own mistakes, biases, and flaws. On the other hand, if we expect AI to do the "right thing," we must confront the challenge of defining what is right. Morality is inherently subjective, varying across different cultures and societies. What one group considers ethical, another might not. This raises complex questions about who gets to set the moral guidelines for AI, and how we navigate the "decision space" where legal frameworks, ethical standards, and what society deems acceptable may point in different directions, leaving us to address conflicting priorities. Examples include issues like capital punishment, gay marriage, or dietary choices, where what is legal or accepted in one country may not be in another, and may have been different in the past or may differ in the future. For example, capital punishment is legal in the United States and China but banned in Europe, reflecting different ethical views on human rights. Gay marriage is accepted in many countries but remains illegal in other parts of the world, and is socially and culturally unacceptable depending on traditions and religious values. Similarly, although prohibiting smoking is ethically encouraged today for health and environmental reasons, smoking was perfectly accepted and legal not so long ago. These differences highlight the challenges of developing AI that must operate within diverse legal, ethical, cultural, and social frameworks, where what is permissible or desirable in one region may be contentious or rejected in another. If what is legally permitted may not always align with ethical principles, and what is ethical may not be socially accepted as "normal" and such understandings diverge in time and across the world, how can AI ever be expected to do the "right" thing?

These questions highlight the significant responsibility we have in shaping AI. The decisions we make now will determine how AI affects society—whether it promotes fairness or exacerbates inequalities. It is crucial to carefully consider who controls AI, how it is used, and how to ensure it benefits everyone.

A Responsible Definition of AI

Despite the many attempts to define AI, there remains no single, universally accepted definition. Different stakeholders, ranging from technologists to policymakers to the media to the general public, often have varying interpretations of what AI is. For some, AI represents advanced algorithms and data processing techniques that mimic aspects of human intelligence, such as learning and problem-solving. For others, AI is seen as a broader concept encompassing any machine or system capable of performing tasks that typically require human intelligence, such as visual perception, speech recognition, decision-making, and language translation.

This ambiguity in defining AI affects how it is perceived, trusted, and understood across different sectors and among various stakeholders. When AI is defined broadly or vaguely, it can create confusion and unrealistic expectations about its capabilities. For instance, viewing AI as a near-magical force that can solve complex problems without human intervention can lead to overconfidence in the technology, resulting in a lack of critical oversight and inflated trust. On the other hand, defining AI too narrowly, focusing only on specific technologies such as machine learning, can obscure the broader societal implications and ethical considerations that come with AI's deployment in various sectors. The lack of a clear, consistent definition also affects trust. When people do not fully understand what AI is or how it works, they may either place undue trust in it or, conversely, become overly skeptical. This polarization can hinder meaningful discussions about AI's role in society and complicate efforts to develop effective regulations and policies. If the public and policymakers do not have a shared understanding of AI, it becomes challenging to establish the necessary safeguards to ensure its responsible use.

These differing definitions can influence how AI is integrated into various industries and how its risks are perceived and managed. For example, in healthcare, where trust is paramount, an AI system perceived merely as a tool to support the decision of medical professionals might be trusted more than one seen as an autonomous decision-maker, even if both systems are fundamentally similar. This can lead to varying levels of acceptance and adoption, affecting the overall impact of AI in different sectors. A responsible approach to defining AI should strive to bridge these gaps, going further than the technology used, to include the social, organizational, and institutional context of that technology. A definition that helps to align understanding across different groups would not only enhance trust but also ensure that discussions around AI are grounded in a shared reality, leading to more effective governance and societal benefits. More than a technology, AI is a socio-technical system, grounded in several disciplines and with broad societal consequences. A multidisciplinary approach to AI can create more effective and ethically sound solutions by integrating diverse expertise and perspectives across various fields. It also presents challenges, such as communication barriers between disciplines, difficulties in integrating different methodologies, and the need for effective coordination and understanding across diverse areas of knowledge.

Key Takeaways and Reflections

The agreement paradox explores the complexities of defining AI and the challenges that arise from the idea that as AI evolves, our ability to clearly define it becomes increasingly elusive. This lack of a consistent definition influences public understanding, research priorities, and policymaking, making it difficult to address AI's impacts comprehensively. This chapter underscores the importance of clear communication and shared frameworks to ensure a common understanding of AI's potential and limitations.

AI's progress comes not from isolated breakthroughs but from the integration of diverse techniques, such as reasoning systems, robotics, and machine learning. Collaboration is crucial—not only between humans and AI but also across disciplines and methodologies. Modular and compositional approaches are particularly valuable, enabling systems to adapt, scale, and remain transparent while addressing societal needs. Ethical considerations are equally important, emphasizing the need for human oversight and the integration of social norms into AI systems. Overreliance on AI for complex decision-making can lead to ethical compromises when systems lack proper judgment or fail to consider context. Embedding ethical principles and ensuring accountability are essential for creating AI that aligns with human values.

In summary, AI's potential lies in striking a balance between ambition and practicality, integrating collaborative and modular approaches while embedding ethical principles into its design. This balance is crucial for developing systems that address societal challenges effectively, transparently, and responsibly.

Ultimately, the way we define AI, or choose not to, will shape how it is developed, trusted, and integrated into our world. The choice, and the responsibility for it, is ours.

3

The Intelligence Paradox

EXPANDING HUMAN CAPABILITIES

IN THE previous chapter, we explored the capabilities of AI and the complexities surrounding its definition, which often lead to varied interpretations and misconceptions. A central issue in this debate is the concept of intelligence itself, the focus of this chapter. Many concerns about AI stem from questions about its intelligence: Is AI truly intelligent? Can it ever match or surpass human intelligence? What would such advancements mean for us? These questions are at the heart of discussions about AI's potential and limitations, driving concerns about the risks it may pose. However, answering these questions starts with understanding the complexity of human intelligence.

Kate Crawford, a leading scholar and author specializing in the social, political, and environmental implications of artificial intelligence, argues that "*AI is neither artificial nor intelligent.*"[1] She asserts that the term *artificial intelligence* is misleading because these systems do not possess humanlike understanding; they simply process data and execute programmed instructions. Additionally, their existence heavily depends on human labor and natural resources, challenging the idea that they are truly "artificial." Many experts, myself included, share Crawford's view that AI systems, despite their advanced processing abilities, lack true comprehension and depend heavily on human input and material resources. Overstating AI's capabilities or anthropomorphizing it can lead to significant misconceptions.

In this chapter, we address the core question: What is intelligence, and how does AI relate to our understanding of it? What AI is or is not, what it can or cannot do, ultimately depends on how we define intelligence and how we believe it can be replicated. Interestingly, in the nineteenth century, Ada Lovelace, the world's first programmer, commented,[2] *"It is desirable to guard against the possibility of exaggerated ideas that might arise as to the powers of the Analytical Engine. In considering any new subject, there is frequently a tendency, first, to overrate what we find to be already interesting or remarkable."*

Back in the 1980s when I started working on AI, an often used definition of AI stated that "Artificial intelligence is whatever machines haven't done yet." This quotation is attributed to Larry Tesler,[3] a computer scientist who worked in the field of human-computer interaction. Tesler observed that once a problem in AI is solved, it ceases to be regarded as AI and is simply seen as another computing task. This concept of AI as a moving target is captured by the central paradox of this chapter:

> ## The Intelligence Paradox
> AI is what AI cannot do.

Consider the example of long division, once a crucial intellectual skill, now easily performed by any calculator. Although no one would classify a calculator as intelligent, the debate continues about whether long division should still be taught in schools. Just because machines can handle tasks like long division or route planning doesn't mean that people are less—or more—intelligent for knowing how to do them as well. The real question is whether relying on machines for such basic tasks is diminishing our intellectual capacity or, conversely, freeing up cognitive space for deeper, more complex thinking. This dilemma highlights the ongoing tension between preserving traditional skills and embracing the cognitive benefits of technological advancement.

My mother, a retired middle-school math teacher, used to believe that mastering long division and calculating square roots by hand were essential skills. However, she later realized that by allowing students to use calculators for these tasks, she could instead spend class time discussing the deeper meaning and importance of calculus and how to approach complex problems systematically. This shift in focus helped students grasp foundational concepts and appreciate how these skills fit into a broader understanding of mathematics. I often reflect on this when I hear concerns about the skills we might be losing as AI takes over tasks like reading maps or writing grammatically correct texts. Just because machines can perform some of our cognitive tasks does not mean we are less intelligent, nor does it mean that machines are intelligent. The processes by which humans and machines operate are fundamentally different, as are the goals. Rather than simply worrying about the skills we may be losing, we should also consider the opportunities we are gaining. What can we achieve with the time and mental space freed up by using tools like calculators, grammar correctors, or navigation apps? Instead of dumbing us down, these tools can enable us to focus on more creative, strategic, and profound aspects of thinking.

THE INTELLIGENCE PARADOX 57

In chapter 1, we discussed how the increasing capabilities of AI are challenging the uniqueness of human intelligence. That is, the more AI advances, the more it highlights the distinct nature of human intelligence, rather than merely replicating it. In this chapter, we further reflect on this effect of AI: the shift on the boundary of what is considered intelligence as more progress is made in the AI field. Is AI the most intelligent entity, or should we see its capabilities as a moving target, coming to appreciate yet other aspects of our own intelligence as AI catches up? We will explore the complex notion of intelligence and how human and machine intelligence can be understood and compared, to conclude that AI is not a continuation of human intelligence but an altogether separate direction of cognitive capabilities. Nevertheless, even if AI is not intelligent in the same way that humans are, this does not mean it is without value—or without risks! In fact, it may pose even more risks, exactly because it operates in ways we do not fully understand. Simply put, the greatest danger arises from our growing reliance on machines to perform functions and make decisions through processes that are not entirely transparent or within our control.

Consider this analogy: Imagine someone introduces a new type of airplane that is supposedly superior to existing models, yet the creators admit they don't fully understand how it flies. Despite assurances of safety, would you trust it? Most of us wouldn't. Yet, this is the situation we face with AI. New applications are being developed that appear to offer significant advancements in areas like image recognition, medical diagnosis, decision support, and even text and image generation, in all of which AI seems to outperform its alternatives. But without the ability to verify or replicate its results, we are left with nothing more than faith, a method far removed from science.

The study of machine intelligence and its relation to human intelligence leads to two different questions: (1) is the human brain a computer? and (2) can machine intelligence replicate all functions of human intelligence? Albeit related, these two questions are very different. Questioning if the brain is a computer shapes our theoretical approach to modeling cognition and influences AI's design principles. In contrast, understanding whether AI can replicate all functions of human intelligence affects AI development and integration into society. I will attempt to answer each of these questions in the following sections.

Human Intelligence

What makes us humans distinct from other animals? Aristotle claimed that it is our ability to reason, make abstractions, and reflect, what he called our "rational soul." For millennia, humans have been indisputably unique in these capacities, but with the advent of AI, some are disputing this uniqueness. Is that rightly so? Are machines really intelligent? And if so, what does that mean for us?

The dictionary defines intelligence as the ability to acquire and apply knowledge and skills. In simple terms, it means being able to understand and learn information, remember it, and use it to handle different situations effectively. Psychologists provide a more detailed definition, suggesting that human intelligence refers to the ability to understand complex ideas, to adapt effectively to the environment, to learn from experience, to engage in various forms of reasoning, and to overcome obstacles by thinking. This comprehensive definition is widely accepted because it acknowledges intelligence as a combination of various cognitive processes, rather than a single ability or skill. Such cognitive processes include the capacity for abstraction, logic, understanding, self-awareness, learning, emotional knowledge, reasoning, planning, creativity, critical thinking, and problem-solving. The Harvard psychologist Howard Gardner, in his 1983 book *Frames of the Mind: The Theory of Multiple Intelligences*, expands on this by suggesting that humans possess multiple kinds of intelligence. Gardner's theory underscores the diverse and multifaceted nature of human intelligence, illustrating that our cognitive abilities are not uniform but vary widely among individuals. According to Gardner, intelligence is not measured solely by traditional IQ tests; instead, it encompasses a range of distinct intelligences, each contributing uniquely to human potential. Gardner identifies multiple forms of intelligence, such as linguistic, logical-mathematical, spatial, musical, bodily-kinesthetic, interpersonal, intrapersonal, and naturalistic intelligence. Each of these represents a different way of processing information and solving problems, reflecting the varied demands of our environments and the complex nature of human existence.

The idea that human intelligence is not a linear progression from less to more, but rather a spectrum of different types, each with its own strengths and applications, mirrors our current understanding of evolution. In biological evolution, complexity and variation are essential for adaptability and survival, challenging the outdated notion that biological evolution follows a straightforward path from "primitive" to "advanced" forms. Likewise, in anthropology, the notion that cultural evolution moves linearly from small, decentralized societies to complex, hierarchical states is now also recognized as both misleading and reductive. Today, these disciplines embrace a more nuanced, branching model that more accurately reflects the diversity and richness of biological and cultural development.

By drawing a parallel between the diversity of intelligence and evolutionary theory, we can gain a deeper understanding of how our cognitive abilities have developed and continue to evolve. Evolution is characterized by a nonlinear progression, with myriad pathways leading to the adaptation and flourishing of organisms in diverse environments. Similarly, human intelligence is not a monolithic construct but a spectrum of capabilities that have evolved to meet various adaptive challenges. This evolutionary perspective on intelligence encourages us to appreciate the unique contributions of each type of intelligence. It challenges the traditional view that values certain cognitive abilities over others and emphasizes the importance of nurturing and developing all forms of intelligence. By recognizing the evolutionary roots and the adaptive significance of diverse types of intelligence, we can foster a more inclusive and holistic approach to education, personal development, and societal progress, as well as to the understanding of artificial intelligence.

Building on this understanding of diverse and branching intelligence, we can explore how intelligence evolved in humans. The social brain hypothesis,[4] proposed by the British anthropologist Robin Dunbar, suggests that human intelligence evolved primarily to navigate large and complex social groups, rather than to solve cognitive problems. Dunbar explains that larger brain sizes, particularly the neocortex, are needed to manage and maintain more extensive social networks. Essentially, as social group sizes increase, the cognitive demands for social bonding, communication, and understanding social hierarchies also rise, necessitating larger brains. This model links brain size directly to social complexity, implying that our cognitive capacities evolved to handle the intricacies of living in large social groups.

To understand this better, we need to consider the different roles of the parts of the brain. Research by Roger Sperry and Michael Gazzaniga in the 1960s, through split-brain studies, revealed that the left hemisphere is primarily responsible for language, analytical thinking, and logical reasoning. These functions are crucial for tasks like communication, problem-solving, and following social rules. On the other hand, the right hemisphere specializes in emotional processing, recognizing faces, and understanding social cues, which are essential for empathy, creativity, and interpreting social interactions. Recent research confirms that the brain has right-left functional asymmetry, but it is more complex than the traditional "right-brain creative, left-brain logical" idea, with both hemispheres interacting dynamically in various cognitive tasks.[5]

Nevertheless, for decades the analogy "the human mind is a computer" has led research in neuroscience, psychology, and also AI. More recently, however, evidence points to the many differences between the human mind and computational systems, revealing that these differences extend beyond the medium (i.e., biology versus silicon) but are far more profound. Moreover, again here, caution is needed when using analogies. As the pioneering neuroscientist Karl Lashley pointed out in 1951: *"Descartes was impressed by the hydraulic figures in the royal gardens, and developed a hydraulic theory of the action of the brain. We have since had telephone theories, electrical field theories and now theories based on computing machines and automatic rudders. I suggest we are more likely to find out about how the brain works by studying the brain itself, and the phenomena of behaviour, than by indulging in far-fetched physical analogies."* Accordingly, evidence suggests that brain functions are best understood as emergent properties, meaning that they arise from the complex interactions of simpler components, like neurons, within

the brain. For instance, György Buzsáki, a prominent neuroscientist, highlights how rhythmic or oscillatory brain activity,
an activity involving the regular, repeating firing of neurons,
leads to complex cognitive functions through the emergence
of patterns that are more than just the sum of their parts.[6]
This view contrasts with the traditional computer analogy,
where the brain is seen as merely responding to inputs and
processing data in a linear way. Buzsáki argues that the computation analogy misses a crucial point: the brain is an active
organ, constantly interacting with and shaping its environment,
influenced by its evolutionary history. The brain is not simply
passively absorbing stimuli and representing them through a
neural code, but rather is actively searching through alternative
possibilities to test various options. That is, the brain does not
represent information: it constructs it. This perspective emphasizes that the brain is not just a passive processor of stimuli but
a dynamic, creative system that actively participates in shaping
its experience and understanding of the world. This emergent
view of the mind's processes explains how high-level, complex
phenomena such as thoughts, emotions, and consciousness
arise from the interactions of simpler neural components,
resulting in properties that are not apparent in the individual
elements alone. On the contrary, although models can replicate
some brain functions, they often struggle with more complex
ones like consciousness. Nevertheless, as computational neuroscience advances, there is growing optimism that models
may increasingly approximate these emergent properties. However, fully replicating them, especially subjective experiences,
remains a significant challenge. As the philosopher John Searle
argues,[7] subjective consciousness may be inherently nonrepli-
cable in artificial systems precisely because of its emergent
nature.

Machine Intelligence

Human intelligence is multifaceted, social, and specialized, and the functions of our brain are complex and possibly emergent. In AI, intelligence is often understood in a much narrower way, and we seem to be stuck with the ladder view of evolution. This perspective is vividly illustrated by the many variations on the well-known image "The March of Progress"[8] that depict a robot as the last figure in a linear progression of human intelligence from ape-like ancestors to modern humans, suggesting the rise of AI and advanced robotics as the next phase in the linear "evolution" of human intelligence and capabilities.

Rather than being aligned in a linear path, machines and the human mind are better understood as two distinct and orthogonal forms of intelligence. Machines excel in processing and executing tasks with incredible speed and precision, often outperforming humans in computational and repetitive tasks. In contrast, the human mind is remarkable for its creativity, emotional depth, and ability to understand complex nuances and context. As to the question of whether machine intelligence can replicate all functions of human intelligence, we can see that currently, AI primarily emphasizes so-called left-brain functions, such as calculation, problem-solving, and rationality, overshadowing other crucial roles of the brain in social and emotional intelligence. AI systems excel at data analysis, logical reasoning, and linguistic processing but struggle with understanding values, emotions, social cues, and empathy. This gap highlights a significant limitation. As we saw in the previous section, current scientific theories stress that human intelligence is fundamentally grounded in social and emotional competencies. Social intelligence allows humans to navigate complex social networks, build relationships, and understand social dynamics, which are crucial for cooperation and survival. However, current AI models ignore, or at best struggle with, right-hemisphere functions such as understanding values and emotions, recognizing social cues, and exhibiting empathy.

The focus of AI on the cognitive aspects of intelligence leaves out many types of intelligence that are crucial for genuine human interaction and to address complex social issues. This is akin to the saying *"If all you have is a hammer, everything looks like a nail."* AI research tends to treat most problems like chess: requiring deep computation within well-defined rules, where progress comes from applying more effort. In contrast, many societal challenges are fundamentally different: the complexity of their solutions comes not from sheer computational effort but from the need to integrate diverse forms of intelligence and different approaches to navigate human contexts, values, and the complexities of knowledge itself.

This brings us to the central paradox of this chapter: AI is what AI cannot do. Even though AI is continually pushing into new territories, it does not cross the frontier of tasks that involve nuanced human judgment, emotional understanding, and social cognition. This highlights the unique and irreplaceable nature of human social cognition and the value of the many branches of human intelligence, which remain essential for complex human interactions and addressing multifaceted social issues.

Much of the discussion about the capabilities of AI is clouded by a lack of understanding of questions such as: What is intelligence? What is computation? What exactly are the functions of the brain? And how does intelligence relate to the ability to interact or affect the world? Among these, **computation** is the easiest to define: it is the process of transforming input into output using a set of well-defined rules, known as an **algorithm**. Simply put, an algorithm is a step-by-step proce-dure, like a recipe, with a finite description. Different types of algorithms exist. A **deterministic algorithm** always produces the same output for the same input by following the same steps every time. In contrast, a **probabilistic algorithm** makes some decisions based on probabilities, so running it multiple times with the same input can result in different outputs, following a probabilistic distribution of possible outcomes. **Stochastic algorithms**, widely used in machine learning and optimization, not only rely on probability but also evolve dynamically over time, modeling uncertainty and continuously refining solu-tions. Finally, a **nondeterministic algorithm** is an abstract concept that can explore multiple possible solutions simul-taneously and select the correct one if it exists, rather than following a fixed sequence of steps. This is useful for solving complex problems where checking each possibility one by one would be too slow or even impossible. For example, in a maze-solving scenario, a nondeterministic algorithm would instantly "know" the correct path instead of trying different routes step by step. Unlike deterministic or probabilistic algorithms, which execute step by step on real-world computers, nondetermin-istic algorithms exist primarily in theoretical computer sci-ence, often associated with nondeterministic Turing machines (NDTMs).[9] Although they cannot be directly implemented on standard computers, the concept helps researchers study prob-lems in optimization, artificial intelligence, and cryptography. Nondeterministic algorithms are fundamental to understand-ing computational complexity, particularly in defining the class of NP (nondeterministic polynomial time) problems,[10] where solutions can be verified efficiently even if finding them may be infeasible for deterministic machines. In machine learning, nondeterminism means that running the same model on the same data can produce different results. This happens due to factors like random initialization of model parameters, stochastic optimization techniques that select random subsets of data during training, and differences in hardware processing. For example, training a neural network for image recognition may yield slightly different accuracy scores each time it is trained, even with the same dataset, because of these random factors. While nondeterminism can make models less predictable, it also helps improve generalization and helps avoid getting stuck in poor solutions.

From the definition of algorithm follows the definition of computation: A function is computable if and only if there exists an algorithm that can compute it. This applies to both deterministic and nondeterministic algorithms; however, nondeterministic algorithms, though theoretically computable, typically require brute force or exponential time on real, deterministic machines. Computation is not limited to computers; our brains also have computational abilities, to process information and solve problems. However, not everything our brains do is computing. In fact, as the computational cognitive scientist Iris van Rooij says, the discussion about the meaning of (artificial) intelligence is not so much about whether the brain can compute (it obviously can) but whether computation is sufficient to describe the cognitive functions of the brain.[11] Understanding the difference between machine and human intelligence requires that we recognize that the brain performs many functions, some of which can be described by computer models, but not all of its activities are purely computational. Additionally, the environment and the body also play crucial roles in how our cognitive processes work.

Current developments in AI have significantly advanced the debate over whether AI can replicate or even surpass human intelligence. These advancements are primarily driven by breakthroughs in machine learning, particularly deep learning, natural language processing, and reinforcement learning. One particular relevant development is large language models (LLMs) like GPT-4, which exhibit remarkable proficiency in generating humanlike text, understanding language, and solving complex problems. This progress is further enhanced by multimodal AI techniques, where systems can process and integrate diverse types of data—such as text, images, and audio. Another relevant development is reinforcement learning, exemplified by systems like AlphaGo and AlphaStar, which have surpassed human performance in complex games by learning strategies through trial and error.

Nevertheless, as we develop increasingly complex AI systems following the narrow understanding of intelligence as discussed earlier, we also uncover other puzzling effects, such as the one described as the Moravec's paradox,[12] which observes that it is not cognitive reasoning but rather the implementation of the sensory-motor and perception skills of humans that requires enormous computational resources. Or, simply put, AI beats humans at chess but it is very clumsy at moving the chess pieces over the board. The same effect is reflected by the fact that the more we automate decision-making, the more human creativity and ingenuity are needed to deal with edge cases and outliers and to give meaning to the insights generated by AI systems.[13]

In summary, not only is human intelligence a complex and poorly understood phenomenon but also the implementation of intelligence within AI systems is a multifaceted topic with implications for the very core of what it means to create entities capable of processing information and making decisions. We will now explore two critical dimensions of AI implementation: the rationality paradigm, and the role of data in shaping AI's capabilities and biases. By examining these aspects, we can better understand the underlying principles that guide AI development and the potential risks and ethical considerations that arise. As AI becomes increasingly integrated into our daily lives, the need for human-guided efforts will be more crucial than ever. While AI uses rational, data-driven methods to process information and make decisions, it misses the (social) nuances that human judgment provides. AI can greatly ease the burden of handling data, but without proper human management of the results and careful control over access, it could lead to significant privacy and security risks.

The Problem with Rational AI

In my book *Responsible Artificial Intelligence*,[14] I spent considerable effort on describing and explaining how AI is traditionally defined within the field of computer science. In computer science, AI is commonly perceived as a system designed for processing information in order to accomplish specific, purposeful tasks. This understanding frames AI as a computational entity, created through human effort, which exhibits reasoning processes and actions reminiscent of human behavior, or behaves in ways that align with our expectations of human cognition or with a rational stance. This classification, along the dimensions *think-act*, and *human-rational*, forms the foundational approach of the most widely used AI textbook: *Artificial Intelligence: A Modern Approach* (AIMA) by Stuart Russell and Peter Norvig. This textbook has educated numerous generations of students, shaping their understanding of AI based on these critical dimensions. The 2021 edition defines AI as "the study and construction of agents that do the right thing" and goes on to loosely define rationality as doing the "right thing"; by their definition, AI would more appropriately be described as "the study and construction of rational agents."[15]

Indeed, the prevailing view sees AI as a rational entity, programmed to maintain coherent beliefs and to prioritize outcomes based on these beliefs. AI applications or agents, as described in AIMA, are constructed to optimize their actions to align with their preferences and understandings, striving to achieve the most favorable outcomes as dictated by their programmed objectives.

Yet, the implications of the impact of AI on society introduce nuances that challenge this paradigm. If AI systems are to interact with us and our environment, and potentially become active participants in our complex world, it is essential to clearly understand the distinction between their formal, rational foundations and the nature of human intelligence. We humans do not usually behave in a formal, rational manner, but we are able to engage with a multitude of possibly conflicting goals and manage a wide range of beliefs that are not always consistent. Our actions are not always "rational," that is mathematically optimal, but are often driven by values such as altruism, fairness, and justice, or by efforts to avoid future regret. Additionally, our decision-making process isn't always about maximizing outcomes; sometimes, we settle for what is good enough rather than the absolute best, reflecting a principle of "satisficing" rather than optimizing.[16]

That is, the rational paradigm not only is insufficient to address the complexity of human behavior but can even be dangerous in dealing with the societal impact of AI. The first danger is the association with human intelligence. Expectations of humanlike intelligence and the idea that machines will behave like humans overlook the profound complexities of human cognition and social interaction, which AI cannot fully replicate, potentially leading to overreliance on or misinterpretation of AI capabilities. Human intelligence, as we discussed earlier in this chapter, is not solely about problem-solving but includes cognitive, emotional, and social facets. As such, a perhaps more fitting definition of AI might be as the capability to utilize knowledge to interact with and adapt to different environments.

The second danger is that these definitions do not fully cover the impact of AI on society. From the perspective of this impact, AI is not just a technology or a computer artifact, but a socio-technical ecosystem that includes people and organizations in many different roles (for example, developer, manufacturer, user, bystander, or policymaker), their interactions, and the processes that organize these interactions and make these systems and their results possible. When the aim is to address the responsible design, development, and use of AI, then it is fundamental to recognize that technology cannot be separated from the socio-technical system of which it is a component. The many guidelines, principles, and strategies for the development and use of AI need to be directed to these socio-technical systems. It is not the AI artifact or application that should be described as ethical, trustworthy, or responsible. Rather, responsibility lies with the people and organizations. In the same vein, ethical AI is not, as some may claim, a way to give machines some kind of "responsibility" for their actions and decisions, and in the process discharge people and organizations of their responsibility. On the contrary, ethical AI gives the people and organizations involved more responsibility and makes them more accountable.

The Problem with Data and Datification

One of the main risks of the current developments in AI is that of datification. This is the process of turning various aspects of our lives and societal activities into data that can be measured, analyzed, and used. With the rise of digital technology, big-data analytics, and internet-connected devices, datification has become increasingly prevalent. In our personal lives, devices like smartphones, fitness trackers, and smart home gadgets constantly track our actions and behaviors. This includes data on how many steps we take, our heart rate, sleep patterns, social media activities, and shopping habits. On a broader scale, societal activities are also tracked through data from public transportation systems, surveillance cameras, social media platforms, and other sources, capturing information on traffic patterns, crime rates, environmental conditions, and public health.

However, the more data we collect, often the less insight we gain. As we drown in a sea of data points, it becomes harder to extract meaningful insights. Models are trained on vast amounts of text data, but adding more data does not necessarily improve their understanding or accuracy. This issue is compounded by the limitations of the currently highly appreciated and widely applied LLMs. These models can generate fluent text but often lack true comprehension and can propagate biases present in the data they were trained on.

Even though opinions diverge on how far we are from the limits of human-generated data, reaching this limit may lead to stagnation in improvements, increased bias, slowed innovation, reliance on outdated information, and higher data costs. However, the alternative, using synthetic data, could be catastrophic for the quality of the content generated by these models, as recent evidence shows.[17] Synthetic data is presented as a promising solution to the demand for large, diverse datasets in areas where access to real data is limited due to privacy concerns, such as healthcare and finance. It enables machine-learning models to be trained on controlled datasets without compromising sensitive information and helps mitigate biases inherent in real-world data, resulting in more balanced and generalizable models.[18] This is particularly useful when real data is scarce or incomplete, helping reduce over-fitting and improve model robustness. However, the utility of synthetic data depends heavily on the quality of the models generating it. If the models are biased or flawed, these issues can propagate, leading to inaccurate results.[19] Furthermore, synthetic data often lacks the complexity of real-world data, which can impair model performance in practical applications. And, as highlighted earlier, overreliance on synthetic data may also result in a significant loss of content quality. All in all, caution is paramount when using synthetic data. Not only may synthetic data present risks similar to real data if not handled responsibly, but also the creation of synthetic data involves making assumptions about the distribution and characteristics of the data, which, if incorrect, can perpetuate biases or introduce new inaccuracies. As always, transparency is crucial in data generation and validation, especially when using AI to generate synthetic data.

Ultimately, and more important, reality is always more nuanced than what data–real or synthetic–can capture. Data points miss the context, subtleties, and deeper meanings of human experiences and societal dynamics. For instance, a higher heart rate as measured by a smartwatch can be due to a person being in love or experiencing heart failure. Without having the wider context, it is impossible to interpret this data in a useful way. The richness of reality often eludes quantification, making it difficult to capture the full spectrum of human life in data. The impact of datification on society is profound. While data-driven insights can improve public services, enhance urban planning, and optimize resource allocation, they can also lead to increased surveillance, social inequality, and a shift in power dynamics. Entities that control vast amounts of data can exert significant influence, potentially leading to a concentration of power.

Data is not a direct reflection of reality; it is constructed. The existence of specific data is shaped by the decisions, choices, and interests of those who create it as well as their capabilities. This means that AI does not access the real world directly but rather a version of it shaped by the data it relies on. Additionally, those who control the data have significant influence over what and how AI "sees" the world. This exacerbates existing inequalities. For instance, if we were to draw a map of the world based on the data that is available and used about each country, the result would be a very distorted map, where the global North, and specifically North America, appears much larger because it accounts for the most data, and the global South regions with little to no dataset usage appear much smaller or are simply not visible.[20] Large tech companies that control most of the data used by AI can use their control to shape societal norms and behaviors. For example, biased search results can reinforce stereotypes, and targeted advertisements can influence consumer behavior and political opinions. Additionally, the dominance of companies like Amazon and Google in their respective markets illustrates how data control can stifle competition and limit consumer choices.

All data is historical. We don't have data about the future, only about the past. Constructing AI models solely based on

training using datasets means that the decisions of those systems will be extrapolations on results from the past. Especially
in dynamic contexts, we know that results from the past are no
guarantee for the future.

How is data constructed? It begins with a decision on which
data is needed, for what, and about what. These decisions necessarily highlight some aspects of reality more than others,
leading by definition to bias. For instance, if we are collecting
data about credit applications, we will probably not include
information about the height or weight of the clients, just as
medical datasets say nothing about the color of your hair. This
means that, later, no decisions will be made based on these
parameters, but also that when that data is reused, it will not be
able to distinguish between these aspects. Many of these decisions are contextual and arbitrary, but others are guided by legal
requirements. For instance, in many European countries, it is
not allowed to include ethnicity in medical datasets, whereas in
the United States this is standard information in almost every
form. When using European data, AI systems will not be able
to distinguish on race, but this does not mean that there is no
race bias in European healthcare!

AI systems struggle with edge cases and outliers in data.
Edge cases refer to situations that occur at the boundaries or
extremes of what a system is designed to handle. For example,
if you're testing a calculator app, an edge case might be entering
the largest possible number the app can handle or dividing by
zero. These are situations that don't happen often but can cause
problems if the system is not prepared for them. Outliers are
data points that are significantly different from most other data
points in a dataset. For example, if you're looking at the ages
of people in a room, and most people are between twenty and
thirty years old, but one person is ninety, that ninety-year-old is

an outlier. Even though we can train algorithms to cover more of these exceptions, at a certain point, the number of resources required for development starts to outweigh the benefit. For example, LLMs are trained over huge amounts of human data but still make mistakes when encountering outliers and edge cases, for example, unfamiliar combinations of words. That is why AI chatbots often say and do things that are absurd. They are not trained to use logic to solve problems; they are trained to predict the next word in a sequence, relying on patterns and past cases, which can lead to mistakes. Handling outliers often requires generalizing beyond a space of training examples. What sets humans apart from machines is our ability to easily apply known principles to unusual situations, making accurate decisions in the moment. We think precisely, evaluating each situation and often getting it right. And we do that because our brain has more complex structures than pattern matching alone. As Gary Marcus, a well-known cognitive scientist and loud critic of the current wave of AI development, explains,[21] *"Even in the simple microcosm of language that is the English past tense system, the brain was a neurosymbolic hybrid: part neural network (for the irregular verbs) part symbolic system (for the regular verbs, including unfamiliar outliers). Symbolic systems have always been good for outliers; neural networks have always struggled with them."*

This goes back to the AI paradox we encountered in chapter 1: The more AI can do, the more it highlights the irreplaceable nature of human intelligence.

What about Large Language Models?

Current developments in AI, particularly in general-purpose models like LLMs, are often portrayed as advancing toward humanlike cognition. These systems, which generate human-like text, have grown increasingly sophisticated, leading some to speculate about the arrival of general artificial intelligence (AGI). AGI will be discussed in more detail in chapter 7, but here we will briefly address the perceived intelligence of LLMs.

LLMs are based on self-supervised learning, allowing them to learn from vast amounts of data without explicit labels, capturing broad patterns applicable to various tasks. The resulting foundational models, though sometimes noisy and imprecise, can serve as a base for various AI applications. The real challenge lies in developing applications that effectively utilize these models' strengths while addressing their limitations. Nevertheless, LLMs represent an incredible advancement, seemingly understanding the world despite being trained solely on language. They have a vast memory capacity, processing more books and articles than a single human could in a thousand lifetimes. This enables LLMs to identify patterns and correlations in data, which can sometimes mimic reasoning about the physical world, but they lack direct experience or true understanding of the world. Moreover, this process differs from actual comprehension. Like a cryptanalyst finding patterns in encrypted messages, LLMs detect relationships and associations in language, giving the illusion of understanding, even though the text may seem as meaningless to them as encrypted messages do to the cryptanalyst. As Paul Lukowicz, a leading ML expert, notes, *"It's misleading to think of 'teaching' an LLM to do something specific; instead, they function as broad, but sometimes flawed, representations of knowledge. The real challenge is to build AI applications that effectively use what LLMs can offer while addressing their limitations."*

The most visible applications of LLMs are chatbots like ChatGPT, Gemini, Llama, or Claude. These systems seem able to hold humanlike conversations due to their pattern-matching abilities over extensive text corpora. However, they still struggle with reasoning, embodied experience, and true contextual understanding, which are essential for accurate real-world reasoning. Their outputs, though often convincing, prioritize plausible text over accuracy, leading some experts to argue that these models produce *bullshit* in the philosophical sense:[22] convincing yet fundamentally unconcerned with truth.[23] Referring to these issues as *hallucinations* misleads the public and policymakers, as these are not bugs but inherent features of how the models are developed. However, it is not the chatbot itself that is the bullshitter. Chatbots, like cars, have no intentions; they simply serve to fulfill human aims. The disregard for truth that users experience in conversations with chatbots should instead be seen as an intentional stance of the corporations behind these technologies, which appear unconcerned with truth for its own sake or its societal impact, but only to the extent that it affects their profits and power.

Another common misconception is equating language proficiency with intelligence. LLMs can generate complex and persuasive text, but this does not imply genuine reasoning or understanding. Given that in the past language has been exclusively used by humans, it is understandable that we now attribute humanlike qualities to these new machine interactions. But these misconceptions can lead to overestimating chatbots' abilities, fostering the belief that they are a major step toward humanlike AI. This confusion arises from conflating language with thought. While chatbots excel in **formal linguistic competence**, that is, the ability to understand and apply language rules, they lack **functional linguistic competence**, which involves using language effectively in real-world contexts that require reasoning, world knowledge, and social understanding. Therefore, and despite their impressive capabilities, LLMs should not be mistaken for true intelligent systems. Their ability to perform well on certain tasks, like passing exams, may reflect memorization rather than genuine understanding. Their well-reported limitations in logical reasoning, consistent world knowledge, and handling complex tasks highlight the significant gap between LLMs and human-like intelligence. Although LLMs are remarkable models of language processing, they are not equipped for tasks requiring deep comprehension, reasoning, or intent. Leading research in neuroscience and linguistics shows that the cognitive functions needed for language comprehension and production involve different brain circuits than those used for listening, reading, or generating fluent sentences.[24] This means that even if LLMs can master vocabulary and grammar, they still are not using language in a humanlike way, and in particular struggle with formal reasoning, acquiring consistent world knowledge, tracking objects or events in long inputs, and generating or interpreting language with intent. This does not mean that LLMs are bad models of language processing, but rather, just as the brain areas that support language cannot do math, solve logic problems, or track the meaning of a story over multiple paragraphs, LLMs also have their limitations on these skills.

In conclusion, although LLMs and other general-purpose AI systems represent significant technological advances, it is crucial to recognize their inherent limitations and the risks associated with overestimating their capabilities. These models, despite their impressive ability to generate humanlike text, do not possess true understanding or reasoning abilities. Misinterpreting language proficiency as cognitive competence can lead to dangerous misconceptions about the potential of AI, particularly in contexts that require genuine understanding and reasoning. As we continue to develop and deploy these systems, it is essential to approach their capabilities with caution and to remain critical of the assumptions we make about their intelligence. Only by maintaining a clear distinction between linguistic output and cognitive function can we responsibly integrate AI into our society and mitigate the risks of overreliance on these powerful yet fundamentally flawed tools.

Key Takeaways and Reflections

The current progress in AI forces us to rethink what we mean by "intelligence," revealing the paradox that intelligence is often redefined by what AI cannot achieve. As AI systems master tasks once thought to require human intellect, like playing chess or generating humanlike text, we move the goalposts, focusing on the abilities AI still lacks. This shifting definition highlights the unique and enduring strengths of human intelligence.

The intelligence paradox shows the complementary nature of AI and human intelligence. AI relies on patterns and correlations in data, excelling at tasks such as identifying anomalies in medical images or optimizing logistics. In contrast, human intelligence thrives on abstraction, causation, and contextual understanding, enabling us to reason creatively, make ethical judgments, and adapt to new situations. For example, although AI can generate a convincing essay on ethics, it cannot meaningfully debate or understand the moral principles behind the text. Similarly, AI struggles with real-world reasoning, such as understanding that a wet road might lead to slippery conditions or grasping the broader implications of societal issues like climate change.

This interplay between AI and humans demonstrates that they work best together. The ability of AI to process vast amounts of data complements human adaptability and social and ethical reasoning, enabling us to address challenges in ways neither could achieve alone. To maximize the benefits of AI, we must ensure that it aligns with our values and enhances, rather than replaces, the irreplaceable qualities of human intelligence.

4

The Justice Paradox

BEYOND BIAS

AS AI technologies become increasingly integrated into various aspects of society, including healthcare, law enforcement, finance, and education, the implications of their deployment grow more significant. While fairness and bias reduction are fundamental aspects of an ethical, responsible, and trustworthy approach to AI, they represent just one piece of the larger concept of justice. Justice in AI relates to the many legitimate concerns about the misuse of AI in the short and medium term, ranging from misinformation and disinformation to deepfakes, and from discrimination and exclusion to the widening of the opportunities gap. Core concepts of justice in AI include fairness, bias, and inclusion. Fairness in AI involves ensuring that systems operate impartially and equitably, offering equal treatment and opportunities to all users. Freedom from bias focuses on eliminating prejudiced outcomes that could treat unfairly or disadvantage specific groups. Inclusion highlights the importance of involving diverse populations in the development and deployment of AI technologies, ensuring that the benefits and advances of

AI are accessible to everyone, regardless of their background or identity.

However, justice in AI is not just about ensuring fairness or reducing bias, as highlighted by our next paradox:

The Justice Paradox
Less bias is not always more justice.

Reducing bias in AI can make outcomes fairer, but it does not automatically lead to more justice or to complete fairness. An AI system might be unbiased yet still lack transparency or accountability, leading users to distrust it due to opaque decision-making processes or a lack of recourse for mistakes. This lack of transparency and accountability undermines fairness and, consequently, justice. Let's pause for a moment to explore the principles of ethical AI:

- **Justice** involves accountability, transparency, and the protection of individual rights, and requires a holistic approach.
- **Accountability** ensures that creators and operators of AI systems are responsible for their impacts and decisions.
- **Transparency** ensures that AI processes and decisions are understandable and accessible to users, promoting trust and ethical use.
- **Protecting individual rights** emphasizes safeguarding privacy, autonomy, and the dignity of individuals, ensuring that AI respects fundamental human rights.

Only when taken as a whole can these principles ensure AI systems that are ethical, trustworthy, and beneficial for all members of society.

Less bias does not automatically mean more fairness, especially if the broader context of structural inequalities is ignored.
Fairness demands that we consider the broader context in
which AI operates. An unbiased AI system might still reinforce existing power imbalances or fail to address the diverse
needs of different groups. For example, even if an AI system is free from gender bias in hiring, it might still favor
candidates from certain educational backgrounds, perpetuating existing social inequalities. True justice involves designing
AI systems that are sensitive to these broader social dynamics and actively promote equity. Efforts to reduce bias can
also sometimes lead to unintended consequences that undermine fairness. For example, an AI system aimed at reducing
bias in loan approvals might inadvertently deny loans to all
high-risk applicants, disproportionately affecting marginalized
communities that already face systemic barriers to financial
access.

Fairness, when rigidly applied, can lead to injustice by ignoring systemic inequalities.[1] Solutionism assumes that fairness
can be engineered, but without addressing deeper disparities,
it often reinforces existing biases. Fair-washing makes mat-
ters worse, presenting flawed systems as neutral while they
perpetuate discrimination. For example, standardized testing
treats all students equally but disadvantages those from under-
privileged backgrounds. AI hiring tools trained on biased
historical data continue to favor privileged groups under the
guise of neutrality. Voter ID laws seem fair but disproportionately exclude marginalized communities. Meritocratic college
admissions prioritize test scores, overlooking socioeconomic
barriers. Even equal vaccine distribution during a pandemic
can be unjust if it ignores areas with higher risks. True justice
requires moving beyond procedural fairness toward equitable

solutions that account for systemic disparities, acknowledging different starting points and addressing the structural barriers that fairness alone fails to fix. Automated systems are perceived as objective and free from the emotional and cognitive biases that can influence human judgment. Since we often associate fairness with consistency and accuracy, the notion that our decisions, and those that affect us, can become fairer by replacing human judgment with automated, data-driven systems is appealing. However, it's important to recognize that these systems are only as fair as the data and algorithms that power them, and therefore often perpetuate or even amplify existing biases. Decision-making processes inherently favor quantifiable features, often overlooking qualitative aspects. This bias toward measurable data, such as using salary or age to assess loan eligibility, tends to overlook "harder-to-measure" qualities like personal history, future prospects, or social and cultural factors.

Justice involves ensuring that not only are rules applied consistently (fairness), but also that the rules themselves are just and promote the overall well-being and equity of society. Justice in AI therefore involves a deeper engagement with ethical, contextual, and societal implications of automated decision-making. AI systems must be designed not only to treat all individuals equitably but also to actively rectify historical injustices and systemic inequities. This requires a multifaceted approach that considers the diverse and dynamic nature of human societies, ensuring that AI technologies are not just impartial but also inclusive and transformative. As Bruce Schneier is quoted as saying, *"If you think technology can solve your [...] problems, then you don't understand the problems and you don't understand the technology."*[2] As we've emphasized throughout this book, AI is not magic; its outcomes are fundamentally shaped by the beliefs, expectations, and decisions of those who build, manage, deploy, and use it. Therefore, to truly grasp the justice of AI's results, it's essential to understand the choices, mechanisms, and stakes that underlie these systems and their decision-making processes.

What Is Justice?

The idea of fairness and justice has deep roots in the human condition and is strongly linked to ethics. That is, in order to understand the idea of justice in AI, we need to start by understanding what justice is, beyond just a textbook definition. The term *justice* has its origin in the Greek language, where it communicates the idea of remaining in your allotted place or role, meaning that a society in which justice has been achieved would be one in which individuals receive what they "deserve." The interpretation of what "deserve" means draws on a variety of fields and philosophical branches including ethics, rationality, law, religion, equity, and fairness. For Plato, justice is about balance and harmony. It represents the right relationship between conflicting aspects within an individual or a community. He defines justice as everyone having and doing what they are responsible for or what belongs to them. In other words, a just person is someone who contributes to society according to their unique abilities and receives what is proportionate to their contribution. Much more recently, John Rawls highlights the foundational role of justice in society. He proposes a social contract argument to show that justice, and especially distributive justice, is a form of fairness: an impartial distribution of goods. Rawls[3] rejects the utilitarian theory of justice,[4] which is based on the maximization of welfare, because of the risk that we might turn out to be someone whose own good is sacrificed for greater benefits for others.

Classical theories, from Plato to Rawls, view justice in opposition to injustice, focusing on addressing various forms of injustice. However, more recent approaches challenge this dualism, offering more nuanced perspectives on justice, in which justice is distinct from virtues like benevolence, charity, prudence, mercy, generosity, and compassion, though they are related. In these approaches, justice centers on fairness and equality, while the other virtues involve kindness and generosity. In particular, ethical theories based on human rights state that the rights established by a society must be protected and given the highest priority. These rights are considered to be ethically correct and valid since a large population endorses them.

The Myth of AI Fairness

Many current debates and governance efforts focus on the question *"Will AI ever be completely fair?"* or variations on it. Taken at face value, the answer to this question is obviously no: "AI will never be completely fair. Nothing ever is." But the issue is also not one of complete fairness, but the need to establish metrics and thresholds to measure and govern fairness that ensure trust in AI systems.

The pursuit of fair AI is currently a lively one, involving many researchers, meetings, and conferences,[5] and it refers to the notion that an algorithm is fair if its results are independent of variables, especially those considered sensitive, such as those traits of individuals that should not correlate with the outcome (such as gender, ethnicity, sexual orientation, disability, etc.). A frequently cited reason for lack of fairness in algorithms is the presence of human bias in the data used to train these systems. For instance, when a job application filtering tool is trained on data describing decisions made by humans in the past, the machine learning algorithm may inadvertently learn to discriminate against women or individuals from certain ethnic backgrounds.[6] This bias can persist even if explicit identifiers like ethnicity or gender are excluded, as the algorithm can still infer these characteristics from indirect cues, such as the applicant's name, address, or word choice. A notable example is Amazon's recruiting AI, which filtered out applications from women because they did not contain the "masculine" language more commonly used by male applicants.

But not all bias is bad; some bias is even necessary. For example, in systems used to prioritize organ transplants, certain biases, such as favoring younger patients with better long-term survival prospects, are intentionally built into the decision-making process. This type of bias reflects ethical considerations and medical priorities rather than unfair discrimination. Eliminating all bias in such a system could lead to less equitable or less effective outcomes, which shows that not all bias is necessarily bad. While striving for fairness in AI, we must recognize that eliminating all bias may not always result in more just outcomes. Currently, we are experiencing an attitude of *"bias against bias,"* which can overlook the nuanced role certain biases play in maintaining context, culture, or diversity within decision-making systems. The real challenge lies in distinguishing harmful biases that perpetuate inequality, from those that reflect legitimate, context-sensitive variations. Achieving true fairness requires addressing both the technical and social dimensions of bias, ensuring AI systems do not inadvertently reinforce existing injustices by oversimplifying complex human realities. Bias is part of our lives partly because we do not have enough cognitive bandwidth to make every decision from ground zero and therefore need to use general-izations, or biases, as a starting point. Without bias, we would not have been able to survive as a species; it helps us select from a myriad of options in our environment. Professionals develop biases from years of experience, allowing quicker, more efficient decisions. For instance, a seasoned firefighter relies on proven procedures during emergencies, and in everyday situations, mental shortcuts, like "common sense," help make fast, practical decisions without extensive analysis. Cultural biases guide smooth social interactions by aligning with social norms and customs. Experts in fields like chess or medicine use pattern recognition biases for rapid, accurate decisions. In high-risk environments, a bias toward caution prevents accidents, as seen with pilots following strict checklists. Biases toward fairness or justice guide ethical decision-making and promote social welfare. Balancing these beneficial biases with critical thinking is essential to avoid negative impacts. Unfortunately, this is a skill that most AI systems lack, showing the importance of human oversight and accountability for the results. Many years ago, I organized an activity at my children's elementary school, helping kids develop fairness standards, roughly mod-eled on Lawrence Kohlberg's stages of moral development. It became clear quite quickly that children ages six to twelve eas-ily understand that fairness comes in many "flavors": if given sweets to divide among all kids of the class, the leading principle was equality (i.e., giving each kid the same amount of sweets). But they also understood and correctly applied the concept of equity: for instance, in deciding that a schoolmate with dyslexia should be given more time to perform a school test. Unfor-tunately, for the average algorithm, commonsense and world knowledge are many light years away from that of a six-year-old, and switching between equity and equality depending on what is the best approach to fairness in a given situation is rarely a feature of algorithmic decision-making. Moreover, nothing is ever 100 percent fair in 100 percent of all situations, and due to complex networked connections, ensuring fairness for one (group) may lead to unfairness for others. What we consider fair often depends not only on the traits of individuals but also on the cultural context we live in. For instance, in the United States, student financial aid is largely need-based, meaning scholarships and grants depend on parental income. This assumes that parents will support their children, often leaving middle-class students ineligible for aid despite financial strain. In contrast, Sweden provides universal student financial aid, including grants and low-interest loans, regardless of parental income. This ensures that students are financially independent and not reliant on their parents' wealth. This reflects different societal and cultural values: American policies assume family financial responsibility, while Sweden prioritizes individual autonomy in higher education.

Suppose now that you are hiring a university professor, and two candidates remain. If one is a man and the other a woman, both equally qualified, the university might choose the woman to improve gender balance. Now, if both are women, but one is national and the other foreign, and the last ten hires were all foreigners, it might seem fair to choose the national candidate to balance nationalities. But what if the national candidate has access to other/more job options, while the foreign candidate has no alternatives? Is it still fair to choose the national candidate? The point is that fairness changes depending on the criteria, and complete fairness is an illusion. We can only say a decision is fair with respect to certain given criteria.

Bias in data is often not the problem; prejudice and discrimination are. Whereas prejudice represents a preconceived judgment or attitude, discrimination is a behavior. In society, discrimination is often enacted through institutional structures and policies and embedded in cultural beliefs and representations, and is thus reflected in any data collected. The focus needs to be on using AI to support interventions aimed at reducing prejudice and discrimination, for example, through education, facilitation of intergroup contact, targeting social norms promoting positive relations between groups, or supporting people identifying their own bias and prejudices. AI bias is more than biased data; there are even biases in the way we approach bias. So, how do we get to justice in AI if bias is already so difficult to solve?

Getting to Justice in AI

Achieving justice in AI requires balancing individual and collective fairness. AI systems must navigate the challenge of treating each person according to their unique circumstances while also addressing broader systemic issues that affect entire groups. This involves weighing the benefits of procedural justice, which emphasizes consistency and transparency, against substantive fairness, which may require different treatments for different individuals to achieve equitable outcomes. A key tension exists between equality and equity: equality involves treating everyone the same way, while equity requires adjusting treatments to account for varying needs and circumstances, ensuring that disadvantaged individuals receive the support they need to achieve similar outcomes. This balancing act also extends to maintaining consistency versus contextual sensitivity, deciding whether AI systems must apply universal principles while remaining adaptable to specific situations and cultural contexts. Additionally, AI developers face the trade-off between impartiality and positive discrimination toward disadvantaged groups. While impartiality demands neutrality and the avoidance of biases, promoting justice sometimes requires proactive measures to lift up marginalized communities, which might be perceived as unfair.

These trade-offs illustrate the complex interplay between justice and fairness, often requiring nuanced and context-specific solutions. Embedding principles of justice into AI necessitates ongoing dialogue between technologists, ethicists, policymakers, and communities. This collaborative effort is essential for developing frameworks that are flexible, transparent, and responsive to society's evolving needs, ensuring that AI systems not only minimize harm but also promote the well-being and dignity of all individuals, particularly those from marginalized and disadvantaged groups.

A foundational result in algorithmic fairness, often referred to as the *impossibility theorem*, states that it's not feasible to simultaneously satisfy all intuitive fairness criteria, such as demographic parity, equalized odds, and predictive rate parity within a single model.[7] This means that efforts to enhance fairness in one aspect may inadvertently lead to trade-offs in another. Although much effort is being put into computational approaches to detect and correct bias, it is important to understand that bias means very different things in statistics and in real life. Statistical bias refers to systematic errors in data collection, analysis, or interpretation that lead to inaccurate estimates or predictions. It arises from flaws in methodology, such as selection methods, measurement errors, or omitted variables, and often distorts research findings unintentionally. On the other hand, social bias involves prejudiced attitudes, beliefs, or behaviors that favor certain groups over others based on characteristics like race, gender, or socioeconomic status.

These biases are rooted in societal norms and cultural contexts, often leading to discrimination and inequality. The relationship between the two becomes evident when statistical bias reflects or amplifies underlying social biases. For instance, if data collection systematically excludes certain demographic groups, it may reveal a social bias in the study's design. Similarly, when algorithms trained on biased data perpetuate unequal treatment, they expose how statistical bias can reinforce social biases. However, statistical bias does not always indicate social bias. In cases where bias arises from technical errors, random sampling issues, or contexts unrelated to social factors, such as studies involving nonhuman phenomena, statistical bias says nothing about social bias. Additionally, if a study focuses on a homogeneous population or lacks social variables, the resulting statistical bias is unlikely to reflect any social prejudice. It is, however, important to recognize that addressing statistical bias alone will not necessarily address underlying social bias. Although correcting statistical methods can improve the accuracy of data and reduce errors, it may not eliminate the deeper societal prejudices and structural inequalities that give rise to social bias. Therefore, efforts to address social bias must go beyond technical fixes and involve broader social, cultural, and policy interventions.[8]

Bias in datasets can arise for many reasons, such as the choice of subjects, omission of relevant variables, changes over time or place, or the way training data is selected. Facial analysis tools and recognition software have raised significant concerns about racial bias in AI. Work by researchers like Joy Buolamwini and Timnit Gebru has shown how deep these biases run and how difficult they are to eliminate. Interestingly, efforts to debias AI can sometimes introduce new biases. For example, Buolamwini and Gebru's creation of a balanced dataset of human faces, called "Pilot Parliaments Benchmark" (PBB) was effective for recognizing race and gender but lacked balance in age or pose. Because of this, this dataset would be unsuitable for testing or training an algorithm to recognize children's faces, highlighting the risks of focusing on certain characteristics while neglecting others influenced by our experiences, time, place, and culture.

Research has shown that it is impossible to simultaneously satisfy all desired properties of fairness, such as calibration across groups and balance in false positives and false negatives. This means that when we calibrate data, we must be prepared to accept higher levels of false positives and false negatives for some groups, and address the significant human and societal impacts that result from that. For instance, in a diagnostic context, a false positive means that a patient is incorrectly diagnosed with a disease they do not have, while a false negative allows a disease to go undetected. Both errors can have profound consequences, both personally and socially. Similarly, being wrongly classified as at high risk of reoffending can lead to severe personal repercussions, such as being held without bail, whereas misclassifying someone as low-risk could endanger society by allowing a real threat to go free.[9] Given these challenges, it is essential to discuss the societal and individual impacts of such errors and to carefully consider the threshold at which we accept algorithmic decisions, acknowledging that they can never be entirely "fair." Although tools like IBM's AI Fairness360[10] and Google's What If Tool[11] provide valuable resources for testing and mitigating bias, they primarily focus on performance rather than justice. This highlights the need for ongoing dialogue about the broader implications of fairness in AI.

AI bias extends far beyond data or the datasets used. It begins with choices about who collects the data, who is involved in selecting or designing algorithms, and who trains and labels the data. Decisions about what data is collected and used and how algorithms are designed and trained significantly affect the fairness of the results. Behind every functioning AI system is a legion of poorly paid, often overlooked human laborers. Books like *Ghost Work* by Mary L. Gray and Siddharth Suri, or *The Atlas of AI* by Kate Crawford, bring attention to these issues, though they often go unnoticed as society focuses on the benefits of AI systems rather than the costs behind their development.

Beyond Fairness

The fact that algorithms, like humans, cannot ever be completely fair does not mean that we should just accept unfair or unjust systems. Improving fairness and overcoming prejudice is partly a matter of understanding how the technology works and a matter of education to ensure not only that technology is developed and used properly but also that fair treatment of those using and being affected by it can be guaranteed. This requires participation and inclusion but still, many stakeholders are not invited to the table, not joining the conversation. The elephant in the room is the huge blind spot we all have about our own blind spots. We correct bias for the bias we are aware of. An inclusive, participatory approach to design and development of AI systems will facilitate a wider scope.

Lack of fairness in AI systems is often also linked to a lack of explanatory capabilities. If the results of the system cannot be easily understood or explained, it is difficult to assess its fairness, especially when the results are not very intuitive. Many of the current tools that evaluate bias and fairness help identify where biases may occur, whether in the data or the algorithms or even in their testing and evaluation. Even if not all AI systems can be fully explainable, it is important to make sure that their decisions are reproducible and the conditions for their use are clear and open to auditing. David Sumpter describes the quest for algorithmic fairness as a game of "whack-a-mole": when you try to solve bias in one place, it appears again somewhere else.[12]

Current AI algorithms are built for accuracy and performance, or for efficiency. Improving the speed of the algorithm, minimizing its computational requirements, and maximizing the accuracy of the results are the mantras that guide current computer science and engineering education. However, these are not the only optimization criteria. When humans and society are at stake, other criteria need be considered. How do you balance safety and privacy? Explainability and energy resources? Autonomy and accuracy? What do you do when you cannot have both? Such moral overload dilemmas are at the core of responsible development and use of AI.[13]

Addressing them requires multidisciplinary development teams and involvement of the humanities and social sciences in software engineering education. It also requires a redefinition of incentives and metrics for what is a "good" system. Doing the right thing, and doing it well, means that we also need to define what is good and for whom.

Finally, it is important to continue efforts to improve algorithms and data, define regulation and standardization, and develop evaluation tools and corrective frameworks. But at the same time, we cannot ignore that no technology is without risk, no action is without risk. It is high time to start the conversation on which AI risks we find acceptable for individuals and for society as a whole, and how we distribute these risks as well as the benefits of AI.

Key Takeaways and Reflections

AI holds great promise for promoting fairness by identifying and reducing biases in decision-making, but the justice paradox reveals a deeper challenge: reducing bias does not always lead to more justice. Biases in AI often reflect those present in the data it is trained on, and efforts to "de-bias" systems may unintentionally obscure the complex realities of social inequities. Addressing these issues requires more than technical fixes: it demands careful consideration of the societal values and contexts in which AI operates.

The justice paradox highlights that fairness is not a universal, one-size-fits-all concept. What is fair in one context may not be in another, and AI systems must grapple with these competing definitions. For example, an AI tool used in hiring may focus on eliminating bias in gender or race but overlook systemic inequalities, such as access to education or professional networks, that shape the applicant pool. As a result, a system designed to be "fair" on paper may still reinforce existing disparities in practice. Achieving justice through AI requires moving beyond technical optimization to engage with the ethical and social dimensions of fairness. This involves recognizing that AI systems cannot resolve complex societal problems in isolation. Instead, they must be part of a broader effort to address structural inequalities, guided by human oversight, diverse perspectives, and transparent accountability. The justice paradox reminds us that fairness is not just about what AI systems do but also about how their use shapes the societies in which we live.

5

The Regulation Paradox

THE ACCOUNTABILITY IMPERATIVE

THE DISCUSSION around AI regulation is as active as the one around AI itself. In general, regulation is seen as the prime way to ensure that AI development is safe, ethical, and aligned with human values. Key reasons to regulate AI include addressing bias and discrimination, ensuring safety to prevent accidents, protecting privacy and data security, holding companies accountable for harm, and preventing misuse in cyberattacks or disinformation. In short, regulation helps protect society while promoting the beneficial use of AI.

Despite many current efforts at guidelines and principles, and even concrete national or regional regulation proposals, most of these initiatives are based on voluntary adherence to the proposed principles. There is, therefore, still significant work to do in the area of regulation. When asked to talk about AI regulation, I often compare the current situation around AI to a car without brakes or seatbelts, driven by someone without a driver's license, going down a road without traffic rules. The only reason we trust ourselves and others to drive (fast) is *because* cars have brakes and there are rules about how vehicles should interact with each other (e.g., on which side of the road vehicles must travel). The fact that regulation on cars has imposed things like brakes, licenses, or seatbelts has not hindered innovation in car engines. On the contrary, such regulations provide the necessary trust and the guarantees that automotive innovation will result in safe vehicles. Similar examples can be seen in other sectors, such as the pharmaceutical or energy sector. The question, therefore, is not whether there is an inherent tension between innovation and regulation, but about defining the *"brakes, licenses, traffic signs, and seatbelts"* needed to guarantee trust and the safe use and development of AI.

The central paradox of AI regulation is, thus, that while a regulation-free environment is envisioned as the most suitable way to encourage innovation and growth by allowing developers and companies the freedom to experiment and push boundaries, it overlooks risks and harms associated with rapidly advancing technologies, and therefore fails to address long-term progress in the field.

The Regulation Paradox
Responsible innovation needs regulation.

The paradox lies in the fact that resistance to regulation often brings attention to the very issues that regulations are meant to fix, such as market failures, ethical problems, or risks to public safety. When industries or people push back against regulation—arguing that it slows innovation, limits freedom, or harms economic growth—they may unknowingly create or reveal problems that show why stronger rules are needed. For example, without proper regulation, risky financial behavior can harm the economy, or environmental damage can worsen without safeguards. The ironic result is that efforts to reduce regulations can end up proving the need for more regulation. This highlights the ongoing struggle between short-term goals, like making a profit or having more freedom, and long-term shared goals, such as public safety, protecting the environment, or keeping the economy stable. This is the essence of the AI regulation paradox. If regulations are not in place, trust in AI could erode, leading to slower adoption or even rejection of the technology. Although it might seem that regulation slows down innovation, it is actually key to building trust and ensuring AI's long-term success. Solving this paradox means finding the right balance—creating rules that manage risks while allowing innovation to thrive.

However, nowadays, the discussion of AI regulation is as hyped as the discussion about AI itself, driven not only by concerns over the potential dangers of an overly powerful technology but also by numerous efforts to resist regulatory oversight. Those who advocate a regulation-free environment are creating a space that leaves risks unaddressed and potential harms unchecked. This, in turn, underscores the necessity for regulation that focuses on those who control AI and not only on the technology itself, as I propose in this chapter. Justice and regulation are related but distinct concepts. Justice, as explained in chapter 4, is about treating people fairly and ethically, based on their rights and what they deserve, and with a focus on equality. Regulation, on the other hand, involves setting rules and guidelines that protect public interests, prevent harm, and create a balanced environment where rights and responsibilities are clearly defined and upheld. The discussion around AI regulation is thus about finding the right balance between justice and the rule of law. Achieving this balance requires careful consideration of ethical principles, legal frameworks, technological capabilities, and societal values. On the one hand, not all laws or regulations are always just, as evidenced in many historical and current examples. For instance, the Jim Crow laws in the United States and apartheid laws in South Africa enforced racial segregation, blatantly violating principles of equality and fairness. Similarly, many countries today still have regulations that criminalize same-sex relationships, restrict gender expression, or deny adoption rights to LGBTQ+ individuals, or enforce regulations that permit industrial activities, such as mining or logging, on indigenous lands without proper consent or consultation with local communities. These examples show how legally enforced regulations may actually perpetuate injustice. On the other hand, laws must be applied fairly and equally, protecting people's rights and maintaining order in society. That is, no one is above the law; everyone, including the government and tech giants, must follow the law. The rule of law also ensures that they, like anyone else, can and must be held accountable if they violate these rules. Examples of how even tech giants and governments are not above the law include the five-billion-dollar fine imposed on Facebook (now Meta) by the U.S. Federal Trade Commission in 2021 for violating user privacy rights, and the European Court of Justice's 2020 ruling that invalidated the E.U.-U.S. Privacy Shield agreement, citing concerns over U.S. government surveillance practices. These cases highlight that both corporations and governments can and must be held accountable for breaking the law.

The discussion around AI regulation is being partially shaped by the lessons learned from the regulation (or lack thereof) of the internet. Although there isn't a direct causal link, the experiences and challenges faced in regulating the internet provide valuable insights into how AI regulation might evolve. Until the mid-1990s, there was no serious attention to the regulation of the internet, which allowed the internet to grow rapidly and innovate without stringent regulatory constraints.[1] The growth of the internet led to a situation where nowadays it is felt to be almost as essential to everyday life as the energy network. However, the lack of early regulation also led to significant issues such as privacy concerns, misinformation, and the concentration of market power among a few tech giants that provide the essential tools to navigate the internet. Similar challenges are now influencing the discussion around AI regulation. Policymakers recognize that, like the internet, AI has the potential for both tremendous benefits and significant risks. This is leading to a growing realization that international cooperation is essential for effective AI governance.

But it is easy to get lost in this complex discussion, so let's try to disentangle all these issues, starting from the core: Does AI need regulation?

Does AI Need Regulation?

AI governance is fundamentally about establishing accountability structures for the development, deployment, and use of AI systems. This challenge extends beyond merely regulating technical solutions. Since AI systems are artifacts, governance should focus on the organizations that develop, implement, deploy, or use AI, rather than on controlling the AI models or applications themselves. Therefore, AI governance is about defining accountability structures for these organizations. Framing governance as merely regulating technology risks overlooking the resulting power imbalance, where private organizations determine and control our ways of working,

living, and interacting, placing us at the mercy of their corporate ambitions.

Those who believe in a non-regulated or loosely regulated
environment for AI argue that overly stringent rules could stifle innovation and economic growth, particularly for startups
developing new AI technologies. They say that being flexible
and able to adapt is crucial, and that developers must be able to
experiment and improve AI systems without being constrained
by rigid regulations. Similar arguments concern ensuring that
local companies can compete globally in the rapidly advancing AI industry, and that too much bureaucracy and delays
in the regulatory process also hinder the deployment of beneficial AI technologies, such as AI in healthcare diagnostics.
Additionally, some believe that encouraging the AI industry to
self-regulate by setting and following its own ethical standards
and best practices, similar to other tech industries, can be a
viable alternative.

But the unchecked expansion of AI technologies by powerful tech corporations presents significant risks today. While
often operating outside the reach of (traditional) democratic
oversight, these entities are shaping our work, lives, and interactions in ways that can lead to profound societal impacts.
The concentration of power within a few large companies risks
creating an uneven playing field, posing threats to privacy,
job security, and democratic processes. Effective regulation
is thus necessary to ensure these technologies are developed
and used responsibly, safeguarding public interests over corporate ambitions.

Furthermore, the potential for AI to enable totalitarian states
to enhance surveillance and control over their populations
cannot be ignored. Authoritarian regimes and rogue actors
could leverage AI to suppress dissent, manipulate information, and infringe upon human rights. Without international

cooperation and stringent global standards, the proliferation of AI in these contexts could exacerbate oppression and reduce freedoms.

So, yes, AI, like any other technology, needs regulation, but there is not a one-size-fits-all solution. Strong political will is needed, and we as citizens can be a part of it. We need to demand "brakes, seatbelts, driver's licenses, and traffic rules" for the AI systems that are influencing our lives and shaping our societies. Governments must step up to create and enforce governance, and take clear positions on the choices they are making, or not making, in the regulation of AI.

What and How to Regulate

Regulation is not just about developing legislation for AI—there are many ways to regulate. Governance is a multifaceted field, in which hard regulation (i.e., legislation enforced by governments) has a place, but also we need to consider soft regulation, self-regulation, social norms and habits, and awareness of the role of AI, as essential components of governance. Much of the current debate around AI regulation stems from a lack of shared understanding of what governance should achieve. Some view regulation as a series of prohibitions on technology development, while others focus on ensuring accountability for the outcomes of AI deployment.

There are many things that can be governed: the technology itself, its applications in concrete fields, the development and deployment processes, the stakeholders involved in these processes, and so on. Take again the example of transportation. There is regulation about the technology itself, the cars, for instance, requiring that cars have brakes, seatbelts, and airbags. But there is also regulation about the use of the cars, requiring that those driving them have driver's licenses. And finally, there is regulation about the interactions (i.e., the traffic rules and traffic lights). Seen from this perspective, as noted earlier, AI systems can be compared to cars without brakes or seatbelts, being driven by someone without a driver's license down a road without traffic lights or rules!

This highlights the complexity that policymakers face when regulating AI, for which they must navigate a matrix of choices, each with its own benefits and trade-offs:

- **Regulate the Technology vs. Regulate Its Applications:** Policymakers can choose between setting broad standards for all AI technologies or focusing on specific applications. Regulating the technology provides a uniform framework for AI development, but it may result in overly broad or complex rules that don't fit every use case.[2] On the other hand, regulating applications allows for tailored guidelines that address the unique risks and ethical concerns of different sectors, like healthcare or finance.[3] However, this approach can lead to inconsistencies across sectors and may struggle to keep pace with rapidly evolving technologies, leading to gaps in regulation across different areas.
- **Formal Legislation vs. Self-regulation:** Formal legislation involves creating laws to ensure consistent compliance and enforcement, offering comprehensive societal protections.[4] However, laws can be slow to adapt to the fast-paced nature of technological advances. Self-regulation, at the other extreme, allows the industry to set its own standards and quickly adapt to changes. However, this can lead to prioritizing the interests of a few over those of the broader public, potentially neglecting important public concerns.[5]

- **Regulation by Prohibition vs. Regulation by Incentives:** Regulation by prohibition involves outright bans or strict restrictions on certain AI practices, setting clear red lines that cannot be crossed.[6] This approach is often used to manage potentially dangerous technologies, ensuring that high-risk activities are curtailed to protect public safety and ethical standards. Conversely, regulation by incentives offers rewards like tax breaks or subsidies for the responsible development of AI.[7] This method encourages and rewards positive behaviors, focusing on promoting compliance with ethical standards and safety protocols rather than relying solely on strict bans.
- **Risk-based vs Rights-based:** Risk-based regulation focuses on assessing and managing potential threats and operational risks within a specific context, aiming to balance various interests, including those of companies. However, this approach may sometimes overlook or downplay the protection of fundamental human rights if the risks are deemed manageable. In contrast, rights-based regulation prioritizes protecting fundamental human rights above all else, ensuring these rights are upheld regardless of the risks involved. Although this approach is essential for safeguarding human dignity, it can face practical challenges in implementation, such as difficulties in applying consistent standards across different contexts.
- **Impact vs. Accountability:** Different regulation approaches will result from focusing on the impact of AI or the accountability of those in charge. An impact-focused approach evaluates AI based on the outcomes of its deployment, allowing regulators to address societal effects and prevent harm, irrespective of fault.[8]

However, this approach might overlook the importance of understanding the intentions and processes behind AI deployment. Emphasizing accountability holds developers and deployers responsible for upholding ethical standards, legal requirements, and public trust.[9] This encourages more cautious and responsible behavior, though it is challenging due to power imbalances between large tech companies and regulators.

Understanding the range of regulatory possibilities is essential for creating an environment where technological advances are pursued responsibly. Effective governance not only enhances innovation and competitiveness but also protects fundamental human rights and ensures societal benefits. This requires addressing every stage of AI development, from design processes and organizational principles to involving all stakeholders in decision-making. Coordinated efforts at national and international levels, combined with embedding ethical considerations into the lifecycle of AI, can ensure that AI drives progress while safeguarding individual values and well-being.

National and global governance initiatives set minimum requirements to ensure a level playing field, providing the public with confidence in the safety and appropriateness of AI systems. However, meeting these minimums is not enough. Governance must also encourage safe, ethical, and innovative AI development that maintains public trust. Companies, developers, users, and policymakers alike must reflect on the principles guiding their actions.

Holding tech firms accountable, both legally and ethically, requires strengthening regulatory bodies with adequate independence and resources, potentially through international

cooperation. Public awareness and education are equally critical. Citizens must understand what AI is, how it affects them,
and their role in shaping its development. For example, just as
public demand for green energy has driven changes in energy
policy, informed citizens can influence AI development to benefit society as a whole. Ultimately, understanding the goals
and impacts of different regulatory approaches is crucial for
effective AI governance. By reflecting on the governance landscape and exploring strategies, we can identify practices that
ensure safety, accountability, and public trust while fostering
innovation. Recognizing that different approaches yield varied
outcomes is key to addressing AI's challenges and building a
future where technology thrives responsibly.

Different Approaches, Different Results

Governance initiatives are often designed along three main
components: which values are core, how accountability is
ensured, and how innovation is addressed. These components are important because they collectively ensure that AI
technologies are developed and used in a responsible manner that is ethical, transparent, and conducive to progress.
Core values provide the ethical foundation, ensuring that AI
respects and protects fundamental rights and societal principles. *Accountability* ensures transparency and trust through
rigorous oversight, risk assessments, and auditing processes,
needed to guarantee that AI systems operate safely and fairly.
Innovation drives the continuous advance of AI technologies
and their uses, promoting technical progress that enables sociopolitical advances, ensuring that AI systems align with social
responsibilities and sustainable development. However, different governance initiatives prioritize different aspects of these

components and select distinct values as foundational. For instance, the E.U. AI Act emphasizes consumer protection as a core value, focusing on safeguarding citizens and ensuring that AI systems are safe and reliable and respect individual freedoms. In contrast, international bodies like the United Nations and UNESCO prioritize broader ethical considerations, aiming to uphold human rights, inclusion, and dignity on a global scale.

In its approach to accountability, the E.U. AI Act also takes a different approach than, for example, the United Nations or UNESCO by prioritizing rigorous risk assessments and auditing processes to ensure that AI systems operate as intended without causing unintended harm, focusing efforts on continuous evaluation, ethical compliance, and interoperability to ensure that AI systems work together seamlessly. In contrast, international agencies such as the United Nations or UNESCO more often emphasize the need to implement global standards for inclusion and equity, ensuring that AI benefits are fairly distributed across all societal groups and upholding transparency and interoperability of AI practices.

In the same way, governance initiatives also diverge on how they aim to support innovation. National or regional strategies often encourage technological advances through tech incentives, to promote the creation and adoption of AI with a focus on national interests, protecting economic goals of their constituencies. Meanwhile, the United Nations and UNESCO emphasize sociopolitical advances, leveraging AI innovation to achieve Sustainable Development Goals (SDGs) and address global issues such as poverty, health, and education.

These different interpretations and priorities lead to varying solutions for AI governance. For example, the E.U. AI Act provides a regulatory framework to ensure safe and trustworthy AI within the European context. International bodies like the United Nations and UNESCO aim to establish global AI ethics and governance standards that can be universally applied, ensuring AI benefits are shared globally. Collaborative efforts by the Council of Europe and OECD seek to create robust AI policies that uphold democratic values and economic stability. Many other initiatives are under way at this time. Notably, the U.N. High-Level Advisory Body on Artificial Intelligence[10] was formed to address the critical need for effective international governance of AI technologies. Its formation is driven by the recognition of both the immense potential and significant risks associated with AI. Its proposals include establishing a global scientific consensus on AI risks and opportunities, harnessing AI to achieve the SDGs, and enhancing international cooperation on AI governance. Additionally, the body provided recommendations on the international governance of AI, fostering inclusive and effective strategies to maximize AI benefits and minimize associated risks. Another development on AI legislation was California's Senate Bill 1047 bill, which aimed to ensure the safe development and deployment of powerful AI systems by mandating safety protocols, including shutdown mechanisms, cybersecurity measures, and reporting requirements, while balancing innovation with public safety through targeted regulations on high-risk AI models. Notably, this proposal also included strong whistleblower protections, allowing employees of AI developers, contractors, and subcontractors to anonymously report safety concerns or legal violations without fear of retaliation, thereby encouraging transparency and accountability in the development of advanced AI models. Table 5.1 gives a short overview of some of the current AI governance proposals, at country, regional, and global levels.

TABLE 5.1. Summary of AI governance approaches globally.

Region	Country/Region Specific	
	Goals	Approach
European Union[a]	Trustworthy AI systems; support innovation	Risk-based legislation: AI Act; strict standards for high-risk AI; innovation incentives
United Kingdom[b]	AI safety; societal benefits	Create AI Safety Institute; minimizing safety risks; rigorous system evaluations; foundational safety research
United States[c,d]		
California[e]	Safe development and deployment of powerful AI systems	Enforcing safety protocols, transparency, and incident reporting; requiring cybersecurity measures and a shutdown mechanism
Canada[f]	Protect citizens; responsible AI; position Canadian firms globally	AI and Data Act (AIDA) legislation: responsible AI development; risk-based approach; consumer protection; human rights laws

China[g]	Protect citizens; support economic growth and national security	Strong government oversight and regulation; secure and stable AI development; ethical AI use
Japan[h]	Societal benefits; economic revitalization	Frameworks for ethical AI use; international cooperation in ethical AI frameworks
India[i]	Inclusive growth; responsible AI use; leveraging AI for social and economic development	Ethical guidelines and regulatory frameworks
Latin America[j]	Unified regional AI strategy; leverage national AI strategies	Adoption of OECD AI principles; various AI governance initiatives
Africa[k]	AI governance aligned with cultural and regional contexts	Data protection laws as foundation; African Union's Continental Strategy on AI (early stages)

Continued on next page

TABLE 5.1. (*continued*)

	Global Initiatives		
Organization	Goals		Approach
UNESCO[l]	Ethical and inclusive AI; focus on sustainability and diversity		Global ethical standards for AI promoting human rights
OECD[m]	Trustworthy AI respecting human rights and democratic values		Guidelines for transparency, accountability, and inclusiveness

[a] https://artificialintelligenceact.eu/

[b] https://www.gov.uk/government/publications/ai-safety-institute-overview/introducing-the-ai-safety-institute

[c] https://www.whitehouse.gov/briefing-room/statements-releases/2024/04/29/biden-harris-administration-announces-key-ai-actions-180-days-following-presidents-landmark-executive-order/

[d] This proposal, announced in 2023 by U.S. Presidential Executive Order, was signed by former President Joe Biden and rescinded by President Donald Trump within hours of his assuming office in 2025.

[e] https://leginfo.legislature.ca.gov/faces/billTextClient.xhtml?bill_id=202320240SB1047

[f] https://www.canada.ca/en/government/system/digital-government/digital-government-innovations/responsible-use-ai.html

[g] https://www.chinadaily.com.cn/a/202001/22/WS5e27a4bfa310128217272e5.html

[h] https://www.meti.go.jp/english/press/2021/0528_002.html

[i] https://www.meity.gov.in/content/artificial-intelligence

[j] https://www.cepal.org/en/news/first-latin-american-artificial-intelligence-index-will-be-presented-eclac

[k] https://ecdpm.org/work/envisioning-africas-ai-governance-landscape-2024

[l] https://en.unesco.org/artificial-intelligence/ethics

[m] https://www.oecd.org/going-digital/ai/principles/

Approaches to Regulation

As mentioned earlier, there are several ways that regulation can support innovation, focusing on the role of incentives, the importance of accountability, and the impact of targeted regulations on research. By examining these areas, we can better understand how regulation, far from being a barrier, can serve as a springboard for AI's continued growth and societal integration.

INCENTIVES

Regulation through incentives offers a proactive and less confrontational way of guiding AI development. When combined with direct legislation and accountability measures, it creates a balanced strategy promoting both innovation and responsibility in AI. However, while regulation by incentives is flexible and promotes innovation, it also faces challenges. Incentives must be designed carefully to avoid being exploited, and relying too heavily on them could create gaps where only minimum standards are met, overlooking broader ethical concerns. To effectively guide AI development, a mix of approaches is needed, combining the flexibility of incentives with the structure of direct regulation. Below are key methods by which regulation through incentives can support responsible innovation in AI:

- **Economic Incentives**: Tax breaks or reduced tariffs encourage companies to develop AI technologies adhering to ethical and safety standards. For instance, companies could receive tax reductions for creating transparent AI systems or investing in research to mitigate job displacement.

- **Subsidies and Grants**: Governments and organizations offer subsidies or grants for projects that advance AI for social good, improve healthcare, or enhance accessibility. These financial supports help foster innovation, especially in areas that may not be immediately profitable but hold high societal value.
- **Recognition and Certification**: Programs certifying AI systems for meeting ethical standards (similar to ISO certifications) incentivize companies to comply with best practices. This certification can enhance their marketability and competitive advantage.
- **Regulatory Sandboxes**: These provide a controlled environment for testing AI technologies under relaxed regulations, allowing innovation while under regulatory oversight. Insights from these tests inform broader regulatory frameworks.

ACCOUNTABILITY

Accountability in AI development is crucial for upholding ethical standards and maintaining public trust. This approach focuses on identifying and holding responsible the individuals and organizations behind AI systems, particularly when these systems lead to harmful outcomes. This aspect of regulation is especially important in addressing power imbalances between large tech companies and regulatory bodies.

- **Transparency and Reporting**: Companies must be transparent about their AI systems, including algorithms and data usage. Mandatory reporting aids regulatory compliance.

- **Audits and Inspections**: Regular independent audits ensure adherence to ethical guidelines, reducing risks like bias and privacy breaches.
- **Legal Liability**: Strengthening legal liability for harm caused by AI ensures that those responsible are held accountable.
- **Power Imbalances**: Large companies, with more resources, can navigate regulations more easily than smaller firms, creating disparities. Big Tech's influence on policymaking through lobbying may lead to industry-favorable regulations, which highlights the need for independent regulators.

Risk

Given the prominent role of risk in the governance discussion, let's look a bit more closely at the concept of risk and risk-based approaches to AI regulation. Current regulatory proposals, notably the AI Act in force in the European Union, use a risk-based approach to classify risks into tiers or levels (such as high, medium, and low risk) based on qualitative and quantitative indicators, which acknowledges uncertainties while providing actionable insights, and helps prioritize which risks need immediate attention and resources. In general, risk is defined as the combination of the likelihood of an event occurring and the severity of its consequences, which is formally represented by:

$$Risk = L \times C$$

where *Likelihood* (L) refers to the probability that a certain event or outcome will occur, and *Consequence* (C) is the severity (or *impact*) of the outcome if the event occurs.

Risk assessment is then the process of determining both the magnitude of potential consequences and the likelihood of those consequences happening. The L x C formula combines qualitative or semi-quantitative ratings of consequence and likelihood to produce a risk score or rating. Essentially, the higher the likelihood and/or the greater the consequences of a potential outcome, the higher the level of risk.

In the AI context, *likelihood* might mean the probability of a system failure, a biased outcome, or a security breach. Estimating these probabilities accurately can be challenging due to several factors. First, AI systems, particularly novel or cutting-edge ones, often lack historical data that could help in estimating probabilities accurately. The complexity and unpredictability of AI behavior further complicate this task. Moreover, the field of AI evolves rapidly, and new vulnerabilities or unforeseen issues can emerge, altering risk landscapes quickly. *Consequences* in AI could range from minor inconveniences to significant societal harm, including loss of privacy, economic disruption, psychologogical or physical harm, or even broader, existential risks. Assessing impact requires subjective judgments, and different stakeholders may perceive the severity of impacts differently based on their perspectives and interests. Impact is also multidimensional, encompassing technical, social, economic, ethical, and legal aspects, which makes quantifying these diverse dimensions in a single metric inherently challenging. Moreover, AI systems may have long-term and far-reaching impacts that are difficult to foresee and measure in the short term.

As an example of a formal risk definition, fatality rate[11] is commonly used to assess the risk of diseases or hazardous activities. However, even in these sectors, risk standards vary substantially. In particular, fatality rate serves as a crucial measure in both aviation and road transportation, but the evaluation and response to risk differ significantly between the two. Aviation's risk evaluation is highly detailed and event-focused, driven by the goal of minimizing catastrophic failures. In contrast, road transportation's risk evaluation is more statistical and trend-oriented, with a higher tolerance for risk due to the nature and volume of road traffic. Moreover, in aviation, there is very low acceptance of fatalities; the goal is often zero deaths because of the catastrophic nature of crashes. In road transportation, there is a higher acceptance of fatalities because accidents are more frequent and widespread, making it harder to achieve zero deaths, with the aim to reduce, not eliminate, fatalities. However, when calculating the risk of self-driving vehicles, the acceptance level of fatalities is expected to be much lower than for regular road transportation, given that the goal of self-driving vehicles is to reduce human error, which causes most accidents. This means that society is less tolerant of any fatalities involving these vehicles, often expecting near-perfect safety performance. As a result, each incident involving a self-driving car is scrutinized heavily, similar to how aviation accidents are analyzed. From these examples, we can infer that even when risk levels are clearly defined and formally calculated, expectations and attitudes toward consequences can be quite different. This highlights the difficulty of applying formal risk calculations to AI. AI applications vary widely in their potential impact, making a one-size-fits-all risk formula impractical or even impossible. The complexity and diversity of these technologies, together with the lack of baselines or historical data, make it difficult to apply a universal risk formula. Instead, risk assessments need to be context-specific, reflecting the potential consequences as seen in sectors like aviation and healthcare. Additionally, a formal approach may result in downplaying or misclassifying risks, and failing to protect fundamental rights adequately, which are non-negotiable and must be upheld regardless of the assessed risk level.

Moreover, when discussing risk in the context of AI, it is crucial to recognize that such seemingly formal equations not only are incomplete but also can be misleading. While providing a framework for understanding risk, both "likelihood" and "consequence" are vague concepts that resist precise quantification. Especially, risk levels do not replace the understanding of *"at risk"* groups, which refer to specific populations or entities vulnerable to certain risks, such as users, developers, organizations, or society at large. Identifying these groups allows for targeted risk mitigation strategies. While risk tiers focus on prioritizing actions, understanding "at risk" groups ensures protection for those most vulnerable, providing a comprehensive approach to risk management. Furthermore, certain AI systems or outcomes might pose a higher level of risk to these at-risk categories than others, necessitating stricter governance measures. However, even in these cases, the circumstances and context remain crucially relevant. For example, risks to individuals versus risks to society cannot be ranked without consideration of context. The National Health Service's COVID-19 contact tracing app in the United Kingdom was designed to monitor virus exposure and alert users to potential infection risks. Although it aimed to protect public health, the app faced criticism over privacy concerns, particularly regarding data storage and access. Critics argued that widespread adoption could set a precedent for government tracking, potentially leading to broader surveillance beyond health purposes.[12] Ranking different risks without considering the context might overlook important nuances. Another example is comparing risks to the economy versus risks to the environment. An AI system that boosts business profits could benefit the economy, but it might also cause risks to the environment, as seen, for instance, in the case of Amazon's AI-powered logistics. By optimizing supply chains and automating warehouses, AI has increased efficiency and profits, benefiting the economy. However, this optimization comes with environmental costs. The increased speed of deliveries leads to higher carbon emissions from transportation. Additionally, the energy consumption of data centers supporting these AI operations is substantial. In regions like Virginia's "data center alley," water usage for cooling these centers has surged, raising sustainability concerns.[13] If economic risks are ranked higher without considering ecological impacts, we might prioritize short-term gains over long-term environmental sustainability.

Given that risk is associated with the capability of the system for autonomous action, it is also important to clarify what autonomy means when referring to machines. Does it refer to a machine's ability to choose its actions, or to set its own goals and motives? To better understand this difference consider, for example, a car navigation system: Although I expect the system to autonomously determine the route to a fast food restaurant, I would not want it to decide whether going to the restaurant is desirable for me or if it should instead direct me to the gym! How we define autonomy will result in vastly different systems with varying impacts. Designing systems under human control, where humans retain decision-making authority, accountability is maintained but the system's efficiency and capabilities may be constrained. If we choose for regulation by design, where regulations are embedded directly into the system or it operates within controlled regulatory environments, risks are reduced but flexibility may be sacrificed. If decision-making is delegated to the AI system, allowing it to handle ethical dilemmas with complex moral reasoning, it raises significant concerns about accountability, as well as questions about the ethics embedded in the system and who determines the appropriateness of those ethical interpretations.

Despite all these challenges, several strategies can improve risk assessment in AI. These include qualitative assessments, such as expert elicitation, scenario analysis, and Delphi methods, to help identify and describe potential risks when quantitative data is lacking. Probabilistic models and simulations can estimate likelihoods and impacts, aiding in understanding potential risk distributions even without precise probabilities. Moreover, the impact of an AI system's failure or misuse can be highly context-dependent, affecting different stakeholders in diverse ways, which complicates the assessment.

Addressing AI risk must include a reflection on the acceptance of risk. Risk is inevitable, but we need to create a shared understanding of what is unacceptable risk. In this debate, a question to address is whether AI applications should be held to different standards than human-driven processes. Holding AI to higher standards can build public trust and ensure that the technology is safer and more reliable. Autonomous systems are expected to consistently follow safety protocols, unlike human processes, which are prone to errors. Ethically, there's a responsibility to minimize harm, and AI has the potential to significantly reduce risks and improve outcomes.

Finally, we should note that a vast array of legal norms regarding risks and harms already exists across global legal systems, making it unnecessary to reinvent these standards from scratch specifically for AI.

Regulation Enables Innovation

An often-heard objection is that increased regulation may lead to less innovation due to higher compliance costs, more complex approval processes, and greater legal uncertainties for developers. The reasoning is that this could discourage start-ups and smaller companies from entering the AI field, reduce investment in research, and slow the pace of technological advancement. Additionally, innovators might hesitate to experiment with novel ideas if they fear regulatory repercussions, potentially limiting creativity and the diversity of AI applications. On the opposite side of the discussion, many experts see regulation as an essential precondition for fostering innovation rather than hindering it. By setting clear guidelines, ensuring safety, and building public trust, regulation provides a stable environment where innovation can thrive. Rather than acting as a barrier, well-crafted regulations serve as a springboard, encouraging investment, experimentation, and the development of new technologies. Just as brakes and safety measures allow cars to go faster, regulation enables innovation to grow responsibly and sustainably.

Nevertheless, as with many emerging issues, the reality is that it remains unclear how AI regulation will ultimately affect innovation. Currently, very few AI-specific regulations are in place, and those that do exist have not been around long enough to provide substantial evidence of their effects. This makes it difficult to draw concrete conclusions about the impact of AI regulation. However, studies from other sectors suggest that well-structured regulations can actually promote innovation by providing legal certainty, setting safety standards, and fostering a more supportive environment for technological progress. Such regulations also enhance user trust and adoption. For example, in the pharmaceutical industry,[14] consistent regulations encourage research investment by providing a stable environment. Similarly, in the financial,[15] environment,[16] energy,[17] fintech,[18] and telecommunications[19] sectors, clear regulations have spurred investment and innovation.

On the other hand, recent evidence suggests that labor regulations can negatively affect technological innovation, particularly when firms face high costs related to regulatory burdens. This is especially pronounced in countries where labor regulations apply to firms of a certain size, causing businesses to avoid crossing size thresholds to evade more stringent rules.[20]

Although concerns about innovation are significant, the need for regulation extends beyond economic concerns. Regulation is crucial for safeguarding public health, ensuring fairness in markets, protecting consumers, and enhancing trust in the safety and quality of products and services. For instance, car safety regulations, such as the mandatory use of brakes and seatbelts, have not hindered innovation in vehicle engines; rather, they have created the trust necessary for people to drive faster and innovate further. Similarly, in AI, the question is not whether regulation stifles innovation but rather what kind of regulatory "brakes and seatbelts" are needed to guarantee safety and trust.

Given the global scope of AI technology, regulation must be addressed on a global scale. Once the need for safeguards such as brakes and seatbelts is recognized in AI governance, it opens the door for broader innovation, not just in AI technology but also in the development of regulatory frameworks, safety mechanisms, and infrastructural advances where multidisciplinary innovation will drive progress in various aspects of AI and its applications.

Can Regulation Keep Up?

As AI technologies evolve rapidly, traditional regulatory frameworks may struggle to keep pace. Static rules and lengthy legislative processes risk failing to address new challenges and opportunities. Adaptive regulation, which allows for flexible frameworks that can evolve with technological advances, focuses on enduring principles such as fairness, privacy, transparency, and inclusion. For example, the European Commission's white paper on AI[21] advocates for a regulatory framework that adapts over time, striking a balance between promoting innovation and ensuring safety and ethics. Examples of adaptive regulation include regulatory sandboxes, which provide a controlled environment where companies can test new AI technologies under the supervision of regulators. This setup allows for real-time adjustments to regulatory requirements based on observed outcomes and provides valuable insights into how regulations might need to be modified.

Critics of adaptive regulation argue that it may lead to regulatory uncertainty, which could deter investment and innovation. They suggest that frequent changes to regulations could create an unstable business environment. Proponents counter that adaptive regulation, when implemented with clear guidelines and stakeholder engagement, can provide the necessary flexibility without sacrificing stability. By involving industry experts, technologists, and other stakeholders in the regulatory process, regulators can ensure that adaptive frameworks are both responsive and reliable.[22]

Global cooperation and standardization are essential for effective AI regulation, given the borderless nature of the technology. Without international collaboration, disparate regulations could lead to fragmentation, where AI systems face different rules in different regions. This could create barriers to innovation and complicate compliance for companies operating across borders. However, achieving global cooperation is challenging due to varying national priorities and approaches to AI regulation.

A Path Forward for AI Regulation

Regulatory efforts in AI need to strike a balance between fostering innovation and ensuring public safety and ethical standards. Although innovation and safety are often seen as opposing forces, it is crucial to recognize that they are not on an equal footing. If the public perceives that safety and ethics are being compromised for the sake of progress, trust in new technologies will erode, leading to resistance, regardless of their potential benefits.

Effective regulation should bridge this gap by ensuring that innovation does not outpace the safeguards necessary to protect society. This requires setting clear safety and ethical standards while also educating and engaging the public about both the benefits and risks of AI technologies. By fostering informed acceptance and support, regulation can create a climate of trust and confidence in AI developments. A phased regulatory approach, similar to pharmaceutical trials, can guide AI technologies from early research to public deployment. This would include, for example, directives for:

- **Prototype Testing**: Ensuring initial models and tools adhere to ethical and safety standards to minimize risks.
- **Pilot Projects**: Testing AI in real-world environments, with careful monitoring to prevent harm.

- **Scale-up Processes**: Addressing broader societal impacts like employment and privacy as technologies are expanded.
- **Continuous Monitoring**: Ensuring ongoing oversight to adapt to new risks and maintain ethical standards after deployment.

This structured approach allows for the exploration of AI's potential while safeguarding the public. Regulatory oversight, akin to that in sectors like pharmaceuticals, can ensure compliance with ethical and safety guidelines throughout each phase of AI development. Engaging stakeholders and maintaining transparency are vital to building the public trust needed to ensure that AI innovations are both socially acceptable and beneficial.

Key Takeaways and Reflections

The regulation paradox highlights the challenge of balancing innovation with accountability in the fast-moving field of AI. Effective regulation is not a barrier to progress but a framework that fosters trust, fairness, and transparency, enabling technological advances to align with societal values. Clear and flexible rules provide the structure needed for responsible innovation, ensuring that AI development addresses ethical concerns without stifling creativity. Regulations also enhance trust by offering accountability and security, giving stakeholders confidence in AI systems and their applications. Moreover, international cooperation is essential to create cohesive standards that keep pace with AI advances and address global challenges.

Far from being a barrier, regulation serves as a springboard for innovation, offering the structure and security needed for technological progress. By providing clear guidelines and fostering trust, regulation allows innovation to flourish responsibly, ensuring that AI development remains not only innovative, fair, and transparent, but also aligned with essential human values.

6

The Power Paradox

CONTROL AND INFLUENCE

TECHNOLOGY IS a multiplier, amplifying both the good and the bad. But if the positive and negative effects of technology grow hand in hand, it will not matter how much better the good gets.[1] The same innovations that improve our lives also open Pandora's boxes of unintended consequences and misuse. Given this duality, we must examine not just the technology itself but also the actors behind its creation and deployment. The real challenge lies in the power held by those who control and shape technology, as their influence often goes unchecked. Do we have enough scrutiny of the individuals and organizations developing these technologies, and of the methods they employ? Without proper oversight, these powerful entities can wield technology in ways that prioritize profit or convenience over safety and ethics, amplifying the risks that come with technological advances and allowing the powerful to shape the future according to their own interests, even if that is at the expense of the greater good.

Contrary to popular narratives that depict AI as a powerful, uncontrollable, existential threat, the real issue around power is our lack of control over the individuals and organizations who develop and exploit AI. The technology itself is not running wild, but the decisions and actions of its creators and owners can lead to significant risks, and at the moment, they are mostly uncontrolled. For example, during a live interview at the ITU AI for Good Summit in May 2024, Sam Altman, CEO of OpenAI, acknowledged that GPT-4 has significant limitations, such as occasionally generating incorrect information or providing misleading answers. He explained that the company is working to fix these problems by using feedback from users to improve the system in future updates–an approach similar to launching an airplane with known faults, trusting that data from midflight incidents will eventually fix the problems. Just as we will not allow a plane to fly without stringent safety checks, relying on users to uncover serious flaws in AI can lead to unforeseen and harmful consequences. Although user feedback is crucial for refining AI systems, this analogy also highlights a deeper ethical concern: the power imbalance between the tech companies developing AI and the general public. When these companies release technology before it is fully tested, they shift the risks onto users, who have little control over how these systems evolve. This dynamic emphasizes the influence and dominance that organizations like OpenAI hold, not just over the technology itself but also over the societal and ethical boundaries within which it operates. The issue is not just about improving AI technology but also about holding these powerful entities accountable to ensure that technology serves society safely and ethically. True progress requires more than ethical principles; it demands enforceable governance and oversight to prevent the exploitation of power for unchecked technological experimentation.

Addressing this issue includes the specification of concrete and enforceable preconditions for the release of AI systems, including requirements for transparency and explainability. Organizations, public or private, deploying AI must be subject to rigorous audits, and those using AI systems should meet stringent expertise standards. Just as pharmaceutical companies undergo extensive testing before releasing new drugs, and drivers must obtain a license before getting behind the wheel, similarly robust frameworks are necessary to ensure that AI is developed and deployed safely and ethically.

This dynamic highlights the power paradox: as we become increasingly reliant on technological systems for everything from communication to decision-making, our dependency gives those who design and control these systems unprecedented power over society. This concentration of power heightens the stakes, making it essential to address not just the tools themselves but also who wields them and to what end.

The Power Paradox
The more AI you get, the less control you have.

The power paradox refers to the fact that those driving technological development in AI wield immense influence over society. With this influence comes the responsibility to act ethically and transparently or, as the saying goes, *"with great power comes great responsibility,"* as popularized by Spider-Man in the Marvel comics and used by historical figures like Winston Churchill, Franklin Roosevelt, and Voltaire. Society expects those in power to justify their actions, ensuring they align with ethical standards. However, major AI tech companies, such as Google, Meta, and OpenAI, have shown a pattern of avoiding accountability for the broader impacts of their technologies, as reported in the 2024 AI Safety Index report.[2]

Further evidence shows that Google DeepMind has achieved groundbreaking advances in AI, particularly in healthcare, but has faced criticism for data privacy violations. In 2017, the company was found to have used patient data without proper consent, violating U.K. data protection laws.[3] Similarly, Meta's powerful AI-driven algorithms on platforms like Facebook and Instagram have been implicated in amplifying harmful content, misinformation, and societal polarization. Despite public outcry and repeated promises to address these issues, Meta's business model continues to prioritize profit and engagement over effectively mitigating these harms. Recent developments have intensified concerns about Meta's commitment to mitigating the spread of misinformation and harmful content on its platforms. In January 2025, Meta announced the termination of its U.S. fact-checking program, opting to replace it with a "Community Notes" system. This shift has been criticized for potentially increasing the dissemination of false information, as it relies on user moderation rather than professional fact-checkers.[4] OpenAI, originally founded with a commitment to safe AI research, has also shifted toward a more profit-driven approach, raising concerns about its ethical obligations. Even though it has published ethical guidelines, OpenAI has been criticized for insufficient transparency and safeguards regarding the responsible use of its AI systems, like ChatGPT. The company's decision to withhold key technical details under the pretext of preventing misuse has conveniently served its commercial interests, restricting independent scrutiny.[5] Legal actions against OpenAI further highlight its avoidance of responsibility. A prominent example is the class-action lawsuit filed by the Authors Guild on behalf of authors such as George

R.R. Martin and Jodi Picoult. These authors accuse OpenAI of illegally using their copyrighted works to train its AI models without permission or compensation, threatening the livelihoods of creators and the broader creative industry.[6] In another case, GitHub faces a lawsuit over its use of open-source code in its AI coding assistant, Copilot, without proper attribution or licensing.[7] Moreover, whistleblower allegations have surfaced, accusing OpenAI management of prioritizing profit over public safety and discouraging employees from raising concerns about potential risks.[8]

A growing number of studies are raising concerns about dependency, or even addiction, to AI chatbots and virtual assistants, with OpenAI itself warning that interacting with an AI voice can foster *"emotional reliance."*[9] This issue is not limited to OpenAI. Platforms like Replika and Character.AI have also seen users forming strong emotional bonds with their AI companions, leading to concerns about overreliance and addiction.[10] The companies developing these technologies hold a significant responsibility here, as their engagement strategies are not unlike the tactics used by social media platforms, encouraging addictive use and fostering user dependency. By creating systems that users may emotionally rely on, these companies could be exploiting human vulnerabilities for profit, much like how addictive substances are marketed.

Despite the clear risks associated with these AI systems, ranging from the generation of harmful content to the deepening of social inequalities and even addiction, tech giants have been slow to implement meaningful measures to address these issues. This pattern of prioritizing rapid expansion and profit over ethical obligations exemplifies how these companies are abusing their power.[11] By failing to establish transparent processes, clear accountability, and concrete safeguards, they allow their influence to grow unchecked, jeopardizing public trust and putting society at risk. Moreover, as discussed in chapter 5, the lack of clear, enforceable frameworks makes it difficult to hold such organizations accountable, leaving room for AI systems to be misused or to perpetuate harm, whether through biases or misinformation. Governments and regulatory bodies must step in to enforce policies that hold these tech companies accountable. Just as the pharmaceutical industry is subject to rigorous testing and approval processes, new technologies like AI should undergo similar scrutiny. Without clear, enforceable frameworks, the unchecked power of a few tech giants will continue to threaten societal well-being, as their technologies may be misused, perpetuate biases, or generate misinformation. This avoidance of responsibility underscores the dangers of concentrated power in the hands of a few entities.

Power and the Powerful

At its core, power represents the capacity for action—the ability to influence events, people, and outcomes. It is an essential force for bringing about change and making a difference in the world. Whether in politics, business, or personal development, power provides the means to effect transformation, create opportunities, and address challenges. It is through the responsible and purposeful use of power that individuals and organizations can shape the future and contribute to the greater good. However, power must be exercised with care, as its misuse can lead to negative consequences, such as oppression, inequality, or unethical behavior. Properly managed, it becomes a tool for positive impact, driving progress and innovation.

In the social sciences, power is typically defined as the ability to control or influence people, resources, or events, and can be understood in several ways, as described by the political scientist and sociologist Mark Haugaard:[12] *"First, there is 'power-to' which is simply the ability to take action or do something. It's the basic capacity for action. Then there is 'power-over' that happens when one person or group makes another do something they wouldn't normally do, often involving some form of control or domination. Finally, there is 'power-with' involving collaboration, where people work together to enhance their collective ability to act."*

In another notable study of power conducted in 1959 by the social psychologists John R.P. French and Bertram Raven, power is divided into five primary sources: coercive power, reward power, legitimate power, referent power, and expert power.[13] These sources provide a framework for understanding the dynamics of power in social interactions. Coercive power relies on the threat of punishment, while reward power is based on the ability to distribute rewards. Legitimate power stems from an official position or role, referent power arises from personal characteristics that command respect, and expert power is derived from specialized knowledge or skills.

Linking Haugaard's dimensions of power with French and Raven's sources of power helps to illuminate the complexities of power dynamics. *"Power-to"* is grounded on expert power and legitimate power, enabling individuals or groups to take action based on their knowledge or positional authority. *"Power-over"* requires coercive power and reward power, leading to the control one party has over another through threats or incentives. *"Power-with"* emphasizes the collaborative nature of referent power arising from respect and shared goals.

In the AI ecosystem, significant power risks arise from the concentration of control within a few dominant companies with the capacity to monopolize knowledge and resources. By hoarding expertise and controlling key technologies (*power-to*), these companies can control innovation and prevent accountability, creating a landscape where they dictate the development and use of AI. This monopolization forces users, industries, and even governments into reliance on their platforms, given that often there is little choice or alternative, enforcing the coercive power companies can wield (*power-over*). Such forced adoption exacerbates the risk of exploitation, especially when companies offer financial incentives or exclusive access to AI systems, persuading stakeholders to adopt AI without adequate ethical scrutiny. Another critical risk is how legitimate power, provided by expertise and capacity to deploy systems, can set biased industry standards, by which companies direct these to suit their commercial interests while neglecting the broader societal impacts. This can lead to the deployment of AI systems that lack proper safeguards, such as privacy protections or bias detection, posing significant harm to the public. Additionally, although collaboration in the AI field holds promise, there is a danger of companies engaging in superficial ethical partnerships such as in the AI alliance[14] (*power-with*). These alliances may appear to address ethical concerns, but without a genuine commitment to change, they serve only to maintain public trust while delaying meaningful regulation. Together, these dynamics show how unchecked power in AI threatens to exploit, manipulate, and harm society if not addressed through robust oversight and regulation.

Moreover, what often goes unexamined is how people obtain and exercise that power. Governance and regulation initiatives are frequently understood as exercises of coercive power, using rules and penalties to enforce compliance. Although this approach can be effective in certain situations, it often overlooks the potential for unintended consequences and the broader spectrum of power dynamics at play. The dynamics of power acquisition and exercise are complex and multifaceted, involving a variety of strategies and methods beyond mere coercion.

Big Tech's Hold on AI Development

The AI ecosystem today is dominated by a few powerful players with vast resources and far-reaching influence. Companies like Google, Amazon, and Microsoft control crucial components of the AI ecosystem, such as data, computational power, and research capacity, creating a structural disadvantage for smaller companies. Startups and research labs rely heavily on these tech giants for essential infrastructure and deployment platforms, further solidifying the giants' dominance. This concentration of power stifles innovation, reduces diversity, and raises accountability concerns. Additionally, these tech giants influence public discourse and regulatory environments by lobbying for favorable regulations and setting ethical standards, maintaining their supremacy and shaping the narrative of AI development. As AI becomes increasingly integrated into critical sectors like healthcare and finance, the lack of transparency within these systems poses significant risks if not addressed.

In ongoing discussions about AI's impact, many have raised concerns about this concentration of power in the hands of too few and too powerful companies, giving them disproportionate control over the rapidly evolving technology. Just as the so-called black box of AI models obscures how decisions are made, the consolidation of AI resources within a handful of companies creates a broader systemic black box, where AI governance is dictated by private entities rather than democratic institutions. The apps we commonly interact with, such as ChatGPT or Midjourney, are just the tip of the iceberg of what can be called the AI stack, which is composed of several interconnected layers, each playing a crucial role in the development, deployment, and management of AI systems, from hardware infrastructure to end-user applications:

- The hardware layer provides the essential connectivity and specialized chips (e.g., CPUs, GPUs) from companies like NVIDIA and Intel, critical for AI performance.
- The data layer covers data collection, processing, and management. Major players like Google and Amazon dominate, while others like Apache Hadoop and Talend focus on processing.
- The algorithm layer involves machine-learning frameworks (e.g., TensorFlow, PyTorch) and pre-trained models from companies like OpenAI and Hugging Face.
- The development layer includes tools like Python and R, supported by organizations like the Python Software Foundation, along with development environments like Jupyter and Visual Studio Code.
- In the training layer, platforms like Google Cloud AI and Amazon SageMaker support model training, while tools like Ray Tune handle hyperparameter tuning.
- The deployment layer focuses on deploying models through platforms like AWS SageMaker, with APIs enabling seamless integration.
- The monitoring and maintenance layer tracks model performance using tools like Arize AI and Amazon SageMaker Model Monitor.

THE POWER PARADOX 137

- The application layer includes end-user applications
like ChatGPT and Google Assistant, where AI models
enhance user experiences.

Figure 6.1 underscores how a small number of corporations control nearly every layer of AI development, from the
underlying hardware infrastructure to end-user applications.
A few players dominate data collection, model training, cloud
services, and deployment platforms, effectively shaping the
entire AI ecosystem. This vertical integration creates significant
barriers to competition, as smaller companies, independent
researchers, and even governments must rely on these firms
for computational resources, datasets, and algorithmic tools. As
a result, AI development is increasingly concentrated within
corporate structures where profit motives outweigh ethical
considerations. At the same time, only a handful of nonprofit
organizations have a significant impact. These include the
Apache Software Foundation, Python Software Foundation,
and the R Foundation, which support open-source projects,
providing essential tools, and frameworks for AI development, helping to ensure that foundational technologies remain
accessible, and promoting innovation and collaboration in the
industry.

The entanglement between AI's core infrastructure and its
applications fosters an opaque ecosystem that limits transparency and public accountability. This is exacerbated by commonly used practices, tools, and frameworks to deal with this
complex environment efficiently. Notably, MLOps (machine-
learning operations) is a set of practices aimed at streamlining
and automating the deployment, monitoring, and management
of machine-learning models in production. Positioned across
multiple layers of the AI stack, MLOps integrates elements

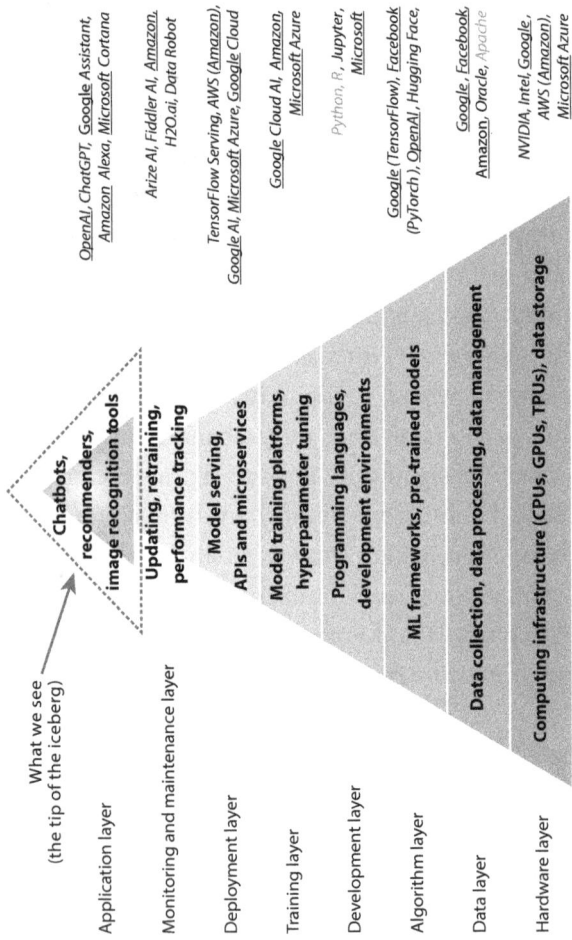

What we see
(the tip of the iceberg)

Layer		Players
Application layer	Chatbots, recommenders, image recognition tools	*OpenAI, ChatGPT, Google Assistant, Amazon Alexa, Microsoft Cortana*
Monitoring and maintenance layer	Updating, retraining, performance tracking	*Arize AI, Fiddler AI, Amazon, H2O.ai, Data Robot*
Deployment layer	Model serving, APIs and microservices	*TensorFlow Serving, AWS (Amazon), Google AI, Microsoft Azure, Google Cloud*
Training layer	Model training platforms, hyperparameter tuning	*Google Cloud AI, Amazon, Microsoft Azure*
Development layer	Programming languages, development environments	*Python, R, Jupyter, Microsoft*
Algorithm layer	ML frameworks, pre-trained models	*Google (TensorFlow), Facebook (PyTorch), OpenAI, Hugging Face,*
Data layer	Data collection, data processing, data management	*Google, Facebook, Amazon, Oracle, Apache*
Hardware layer	Computing infrastructure (CPUs, GPUs, TPUs), data storage	*NVIDIA, Intel, Google, AWS (Amazon), Microsoft Azure*

FIGURE 6.1. The AI stack (underlining indicates big tech organizations that operate at different layers and therefore can ignore horizontal agreements; gray indicates nonprofit, mostly open-source, players)

from the data layer, training layer, deployment layer, and monitoring and maintenance layer to create a cohesive system for managing the entire machine-learning lifecycle. However, the automation of processes can make it difficult to keep track and understand how the different models make decisions and how they are maintained. This lack of transparency can lead to serious problems, especially in areas where accountability and explainability are crucial, such as healthcare, finance, and legal systems. Without clear documentation and transparent workflows, MLOps can create a "black box" effect, where the inner workings of AI systems are hidden from view. This can undermine trust, hinder troubleshooting, and make it hard to ensure that models are working correctly and fairly. The opacity inherent in the current data-driven paradigm of AI built according to this stack can hinder trust and accountability, as stakeholders cannot clearly see how decisions are made or trace the origins of a model's behavior. In critical applications, this can lead to dangerous outcomes, legal liability, and ethical concerns. Therefore, it is crucial to devise alternative architectures and models with a strong emphasis on transparency, accountability, and participation, as we will discuss further in chapter 8.

Figure 6.1 may give the impression of a clean, clear-cut separation between layers and concerns, but the concentration of power across the AI stack makes it better compared to a mass of Play-Doh. Each layer of the AI stack, initially distinct, begins to blur as those controlling them exert their power, much like how different colors of Play-Doh merge into a single, indistinct mass. Once merged, it becomes nearly impossible to separate the colors, or in the case of the AI stack, it becomes increasingly difficult to trace decisions and actions back to their origins. This highlights the urgent need for greater transparency and replicability in AI systems.

In the context of the current AI paradigm, based on building larger and larger models, there is virtually no AI without Big Tech. Only a few companies have the resources to create and deploy large-scale AI models, allowing them to make crucial decisions that affect society. These companies are driven by profit and shareholder returns, not necessarily the common good. Yet, paradoxically, it is this very accumulation of power that imposes an ever-increasing obligation to justify and explain their decisions and to be open to scrutiny by others.

The concept of compositional AI is gaining momentum as an alternative to the monolithic architectures that dominate today's AI systems. By embracing modularity, AI can be designed as a collection of smaller, independent components—each developed, tested, and optimized separately. This approach enhances flexibility, transparency, and adaptability while making AI systems easier to manage and refine. Compositional AI also emphasizes layered and hybrid architectures. Layered systems structure AI into different levels of abstraction, allowing for both high-level strategic oversight and detailed technical execution. Hybrid systems, on the other hand, integrate AI with human expertise, ensuring that AI-driven decisions remain contextually relevant and ethically grounded. This is particularly valuable in uncertain or ethically sensitive domains, where human judgment remains indispensable. Additionally, a modular AI framework fosters collaboration beyond disciplines, enabling seamless integration with human expertise and other technological advances. By shifting away from rigid, all-in-one AI models, Compositional AI paves the way for more adaptable, responsible, and human-centered AI development.

Trust and Trustability

Trust in AI is fundamentally a human, societal question and should not be discussed without considering how it relates to humans. People are involved at all points in the lifecycle of an AI system: from the system's decision-making to its design, development, governance, and social impact. Therefore, it is important to distinguish between trusting that AI as a tool will work as expected (that is, that it will operate reliably and predictably) and trusting that the people and organization involved in the development of AI systems understand their responsibilities and will fulfill their role in a reliable manner. AI systems operate in diverse environments where people live and work, often in unpredictable conditions that change over time and space, where situations we have never encountered before arise. Trusting that the AI system will not exhibit undesirable behavior in such environments or, at the very least, that the effects of such behavior can be kept below some acceptable threshold, is critical for AI adoption. To build trust in AI decisions, we need to understand how the system arrives at its conclusions and recommendations. Trust in the system can be enhanced through different means, such as transparency about the motives, goals, and vision of those behind it, clear explanations of its functioning, or education to enable stakeholders to better understand its capabilities and limitations.

Algorithmic transparency is the principle that the factors influencing algorithmic decisions should be visible or transparent to the people who use, regulate, or are affected by them. In a strict sense, algorithmic transparency is a red herring, simply solved by making code and data open for inspection. However, this "solution" does not lead to trust: not only does it risk violating intellectual property, security, and business models of those developing the algorithms, but most users would not find the code comprehensible.

Alternatively, we can turn to explanation as a prerequisite for trusting AI systems. Explanation helps reduce the opacity of the system, supports understanding of its behavior and limitations, and provides clarity when things go wrong. Post-mortem explanations, akin to the black boxes used in aviation, help investigators determine what went wrong. Yet, AI systems are often held to a higher standard than human decision-making, as excuses like being distracted or confused do not apply to machines. Another reason for the need for explanations is that machines lack accountability, which drives the need for proof or certification of their reasoning capabilities, or at least a guarantee about the scope of the system's decisions.

Explanations can be divided into two types: faithful and plausible. Faithful explanations focus on accurately showing how an AI system makes its decisions. They aim to provide a true reflection of the model's reasoning so it can be verified and understood clearly. These explanations can be built into the model from the start or added afterward by analyzing its behavior. Plausible explanations, on the other hand, focus on making the AI's decisions understandable and useful to people. They may not show the exact inner workings of the model but rather aim to present the decision in a way that is relevant, clear, and socially meaningful. Both types of explanations are important—faithful explanations ensure accuracy and trustworthiness, while plausible explanations help people make sense of the system's decisions in real-world contexts. Explanations should also be contrastive (that is, answering the question "Why X instead of Y?") and selective (focusing on relevant factors).[15]

Education also plays a crucial role in fostering trust, as a well-informed public is better equipped to engage with and adopt AI systems. Misconceptions about AI's capabilities, such as the belief that AI is infallible or, conversely, that it is too unreliable, can spread easily, eroding trust in the actual potential of these systems. Without continuous efforts to educate the public, AI advances will lead to fear, confusion, or unrealistic expectations. Therefore, educational initiatives are essential not only for users but also for policymakers, developers, and society at large to understand the benefits and limitations of AI. A comprehensive educational approach should address not just how AI works, but also the ethical considerations and risks, and how it can enhance human decision-making. Collaboration across academia, industry, and government is vital to establish the necessary knowledge to build trust and reassure consumers that their best interests are prioritized, in both design and implementation.

Trust, explainability, and transparency are deeply interrelated concepts that influence one another. Interestingly, trust can exist even in the absence of complete explanations, as trust is context-dependent. For example, I might trust an automatic checkout system at a supermarket based on prior positive experiences, even if I don't fully understand how it processes payments. However, in more critical situations, such as a medical diagnosis, the stakes are much higher, and I would require a detailed explanation of how the AI reached its conclusions before placing my trust in the system. This demonstrates that trust can sometimes be built on perceived reliability or familiarity, but in other cases, trust is contingent on transparency and explanation to ensure accountability and informed decision-making. In all cases, measuring trust in AI systems is a complex and nuanced task that goes beyond simply tracking system performance. Trust is inherently subjective and varies between individuals and contexts. It requires human-based metrics, both qualitative and quantitative, that assess not only the system's reliability but also the quality of the interaction from the user's perspective. These metrics might include immediate human performance, user satisfaction, and the user's ability to communicate needs effectively. Additionally, trust metrics should consider the user's willingness to explore alternative solutions, the transparency of the system's decision-making process, and the long-term effects on user skills and decision-making abilities. For instance, repeated positive experiences with an AI system may increase a user's trust over time, while a single failure without adequate explanation could have the opposite effect. Trust metrics must be dynamic and capable of adapting to the evolving relationship between humans and AI systems.

Power and Democracy

As we have seen, the AI ecosystem is heavily dependent on a few large organizations, which leads to a perpetual power cycle: as these companies grow and consolidate their influence, public skepticism and wariness increase. In response to this distrust, companies may prioritize short-term competitive advantage and investments over ethical standards and responsible governance, believing that such actions will help maintain their market position and profitability. In these efforts, they might cut corners in safety protocols, data privacy, and fairness in AI systems to reduce costs and accelerate development. They may also exploit loopholes in existing regulations or lobby against new policies that would impose stricter ethical and operational standards. Moreover, large organizations often have significant resources to influence policy and regulatory decisions through lobbying efforts, shaping regulations to favor their interests or delaying the implementation of stringent standards. Then, this focus on immediate gains further fuels public distrust and hinders the development of a more inclusive and ethical AI landscape.

Concentrating power in a few big tech companies worsens existing problems in democracy, such as the erosion of trust in institutions, misinformation, and unequal participation. People are losing faith in government, media, and other democratic institutions due to perceived inefficiencies and a sense that these institutions no longer work in their best interest. This loss of trust weakens democracy, as institutions depend on public confidence to function effectively. At the same time, governments struggle to regulate big tech companies effectively. These companies wield enormous influence over public discourse and political outcomes while also making public institutions dependent on their services. For example, many government agencies use cloud services and communication platforms from tech giants, which increases their control over how public institutions operate and communicate with citizens.

However, besides examining the concentration of power in the hands of Big Tech, it is equally important to address how AI is being leveraged by authoritarian regimes for surveillance and control, further intensifying the conversation about the intersection of technology and governance. This dynamic highlights the broader implications of AI as a tool for reinforcing power structures, both corporate and political. Just as Big Tech companies have amassed vast power through AI-driven innovations, authoritarian governments are leveraging similar technologies to tighten their grip on citizens, underscoring the dual-edged nature of AI's influence on global power structures.[16] For instance, in China, AI technologies such as facial recognition and social credit systems[17] are deployed to monitor and control large segments of the population, including minority groups like the Uyghurs.[18] Despite China's being a signatory to international agreements like UNESCO's AI ethics guidelines, its use of AI serves to enhance state surveillance, not individual empowerment. The increase in surveillance following public protests further illustrates the regime's growing reliance on AI to maintain control, creating what scholars describe as an "*AI-tocracy*": a state where political power and AI innovation reinforce one another.[19] Similarly, Russia has employed AI, particularly facial recognition technology, to monitor political opponents and activists.[20] Despite international agreements, the lack of oversight or regulation in these AI applications has led to widespread privacy violations and restrictions on freedom of speech. In Iran, the government has significantly ramped up the use of AI, particularly in enforcing strict social and religious codes. For instance, AI-powered facial recognition technology is being deployed to identify women who fail to comply with hijab mandates. In 2023, reports indicated that millions of women received warnings that their vehicles could be seized if surveillance cameras caught them without hijabs.[21] This use of AI has been criticized as a tool of "gendered repression" designed to suppress women's rights movements and broader dissent against the regime. In Israel, AI is heavily integrated into military and surveillance operations, particularly in the occupied Palestinian territories. Facial recognition systems are employed to profile and track Palestinians, with soldiers instructed to add photos to a central database. These tools, used at checkpoints and across the West Bank, enable real-time monitoring and control of movement. Critics argue that this creates an "automated apartheid," subjecting Palestinians to constant surveillance and restricted mobility, while also supporting targeted military actions.[22] In addition, Iran has used AI-enhanced cyber operations to spread disinformation and attack political targets as part of its broader geopolitical strategy, including cyberattacks against Israel.[23]

The power of AI in these contexts highlights a troubling aspect of the global AI landscape: while Big Tech concentrates technological and economic power, authoritarian governments are wielding AI as a tool for political control. This intersection of technology and authoritarianism calls for stronger global accountability mechanisms to ensure that AI development aligns with ethical standards, not just on paper but in practice. Not only autocratic regimes but also democratic countries have been criticized for their use of AI technologies in ways that can potentially undermine privacy and civil liberties.[24] In the name of national security or public safety, democratic governments have sometimes employed AI surveillance tools that raise ethical concerns. For example, in the United States and the United Kingdom, facial recognition technologies have been implemented by law enforcement to monitor public spaces, raising concerns about privacy violations and potential misuse of data without sufficient oversight. The widespread use of AI in predictive policing has also been criticized for reinforcing racial biases by relying on historical crime data that disproportionately targets minority communities. During the COVID-19 pandemic, several democratic governments, including those in Europe, introduced AI-driven contact tracing and surveillance systems. While these systems aimed to curb the virus's spread, there was significant concern about how such data could be misused for other purposes beyond health monitoring, such as tracking political dissidents or activists.[25]

Misinformation is another growing threat to democracies and autocracies alike, where social media platforms allow false or misleading information to spread with unprecedented speed and reach. Unlike traditional media, which often follow editorial standards, digital platforms prioritize engagement, amplifying sensational or divisive content that can quickly shape public opinion. The rapid dissemination of misinformation influences key aspects of society, including voting behavior, policy debates, and public trust in institutions. When people are bombarded with false narratives, it becomes increasingly difficult to distinguish between fact and fiction, eroding the foundation of informed decision-making on which democracy depends. This, in turn, weakens public confidence in electoral processes, governance, and even basic civic responsibilities. Moreover, misinformation is not a neutral force; it is often intentionally spread to achieve political or ideological goals. Foreign interference, disinformation campaigns, and politically motivated actors exploit the virality of false information to polarize societies and sow confusion. In such an environment, truth becomes a contested space, and citizens are left questioning the integrity of democratic institutions. As misinformation continues to proliferate, addressing it through stronger content moderation, media literacy education, and transparent information ecosystems becomes crucial to preserving democratic integrity.

Participation in democratic processes is another critical issue that is exacerbated by the unequal access to resources and education. Marginalized communities, whether due to socioeconomic status, geographic location, or lack of educational opportunities, are often excluded from full participation in civic life. These groups may have limited access to reliable information, the internet, or the skills necessary to engage meaningfully in political discussions, which limits their ability to influence decisions that directly affect them. For instance, rural areas, low-income communities, and minority groups are often underrepresented in political processes, leading to policies that overlook their needs or exacerbate existing inequalities. This unequal participation weakens the democratic principle of fair representation. When certain voices are systematically left out of the decision-making process, policies tend to favor the interests of more privileged groups. Over time, this exclusion fuels disenfranchisement and disillusionment with the political system, further widening the gap between those who hold power and those who do not. For democracy to thrive, it is essential that every citizen has the opportunity and means to participate fully, and this requires addressing the barriers that prevent equitable participation.

AI technologies are worsening both the spread of misinformation and the barriers to democratic participation. Algorithms on social media platforms can be weaponized to target specific groups with personalized, often misleading, content. These AI-driven systems analyze vast amounts of data to identify individuals' preferences, biases, and vulnerabilities, and then deliver tailored messages that reinforce existing beliefs, amplify misinformation, or manipulate opinions—often without users being aware. This micro-targeting, combined with a lack of transparency about how algorithms operate, undermines the possibility of a shared, fact-based understanding of societal issues, further polarizing the public. Additionally, AI exacerbates unequal participation by reinforcing existing power imbalances. Those with access to advanced technologies and vast datasets have a disproportionate influence over public discourse and political outcomes. Tech companies and political actors who can afford sophisticated AI tools can manipulate public opinion more effectively than marginalized groups, further entrenching inequality. Meanwhile, the voices of communities with fewer resources or less digital literacy are drowned out, making it even harder for them to participate in democratic processes. Such concentration of power in the hands of a few large tech companies limits transparency and accountability. These corporations control not only the flow of information but also the tools that shape public opinion. Their lobbying efforts often influence regulations in ways that benefit their own interests rather than the broader public good, making it difficult to address the deepening inequities in both information access and political participation.

Expanding access to education, ensuring equal representation, and creating opportunities for marginalized communities to participate fully in democratic processes are essential to safeguarding democracy in the face of growing technological influence. This requires strong democratic institutions, combating misinformation, and ensuring equal participation for all. AI can significantly enhance the strength of institutions and regulation by improving efficiency, transparency, and accountability. In regulatory processes, AI systems can automate monitoring and compliance checks, quickly analyzing large datasets to identify potential violations in areas like finance, environmental protection, and data privacy. This allows regulators to respond more swiftly and accurately to infractions, ensuring better enforcement of laws. Moreover, AI can improve transparency by making government processes more open and accessible, helping institutions track performance, identify inefficiencies, and increase public trust. By providing tools for better data analysis and decision-making, AI can support institutions in creating fairer, more responsive, and accountable regulatory frameworks.

Finally, we must avoid treating AI development as a competitive "arms race" with winners and losers. True progress comes from long-term, responsible research that considers societal impacts, improves governance, and focuses on sustainability and human rights. Just as innovations in green energy and fair trade have set higher standards, we urgently need responsible AI innovations that prioritize public trust and ethical alternatives.

Geopolitical Tensions: Cooperation as True Competitive Advantage

If there were any doubts about the reality of the power paradox– that is, that increased AI capability can simultaneously erode control–the current geopolitical tensions offer a compelling demonstration. The current attitude appears to be that nations must engage in intense competition, driven by the belief that AI leadership translates directly into strategic advantage. Yet, paradoxically, this competitive drive heightens instability, exacerbates inequalities, and challenges traditional geopolitical balances, turning AI into both a tool of power and a source of vulnerability.

Current geopolitical tensions, notably the competition between the United States and China, illustrate how AI fuels a technological arms race. Both countries prioritize rapid AI advancement, often at the expense of careful oversight, sustainability, and ethical considerations. From 2022 on, the United States has implemented and expanded export controls on semiconductor technologies and related equipment to China, aiming to limit China's access to advanced technologies and hinder its military modernization efforts, demonstrating that AI has become a central theater for economic and geopolitical influence. However, instead of ensuring dominance, these restrictions often push countries to speed up their own technology development, which can undermine global stability and make international cooperation more difficult. For example, recent developments like China's rapid advances in AI through models such as DeepSeek illustrate how nations respond by swiftly developing their own powerful AI tools, further complicating international oversight and collaboration.

Moreover, authoritarian regimes use AI not merely as economic leverage but explicitly as instruments of control, surveillance, and cyberwarfare. Examples include Iran's use of AI-enhanced cyber-operations to launch disinformation campaigns and cyberattacks targeting geopolitical rivals like Israel, or Russia's deployment of AI-driven propaganda and deepfake technologies to influence public opinion and destabilize democratic processes.[26] Similarly, China's significant advances in AI-driven surveillance technologies raise concerns about human rights and shift global norms concerning privacy and autonomy.[27] These developments highlight the role of AI in enhancing authoritarian power structures, complicating efforts to establish universal ethical standards and governance frameworks.

Democratic nations also face significant challenges related to AI deployment. For instance, the use of facial recognition and predictive policing systems in countries such as the United States and the United Kingdom has generated controversy due to concerns over civil liberties, racial bias, and intrusive state surveillance,[28] illustrating how even democratic states, when driven by geopolitical competition or domestic security pressures, can compromise transparency, accountability, and ethical standards to achieve perceived technological superiority.

But a competitive approach to AI development, driven by geopolitical rivalry, risks becoming a harmful race to the bottom, prioritizing speed over ethics and transparency. Paradoxically, the real advantage comes not from competition but from international cooperation and shared governance. Without collaboration, AI will create mistrust, fragmented rules, and greater security risks. Additionally, the global AI governance landscape itself risks being dominated by a few powerful nations and large technology companies, often excluding critical perspectives from the global South. This exclusion reinforces existing inequalities and diminishes the legitimacy and effectiveness of international AI governance initiatives.

Moreover, current geopolitical competition in AI can limit international access to critical research and infrastructure, further fragmenting global efforts in AI governance. This rivalry underscores the urgent need for greater international cooperation in AI research and innovation, without which nations risk deepening mistrust, isolating themselves technologically, and potentially weakening their overall security and resilience.[29]

The power paradox offers crucial insights into addressing current geopolitical tensions around AI. Recognizing that increased AI capabilities can paradoxically reduce control means understanding that as AI becomes more powerful and advanced, it also becomes more complex and harder to manage. Instead of giving nations clear control or guaranteed dominance, advanced AI may introduce unexpected challenges and risks, making outcomes harder to predict and manage effectively. This reality suggests that nations should rethink competitive approaches, moving instead toward international collaboration, guided by shared governance frameworks, transparency, and commitment to human rights. Such cooperation can mitigate tensions, build trust, and foster global stability.

Ultimately, real power in the age of AI comes not from unilateral technological dominance but from inclusive, cooperative governance. By embracing collaborative strategies over rivalry, countries can effectively address geopolitical conflicts, ensure sustainable progress, and align AI advances with global well-being.

The Way Forward: Power to the People?

Even though AI can be seen as just one more step in the trend toward automation, it carries deeper implications than other technologies because it touches on core human concepts such as intelligence and autonomy. This raises critical questions about how AI will affect our jobs and lives. Why do we perceive AI and robotics as more dangerous than earlier technologies, fearing that they may take over our jobs rather than support us in some tasks? One answer relates to the speed of change. Technological innovation is not new, but in the past two centuries, its pace has increased dramatically. Since the nineteenth century, advances in productivity, particularly in agriculture, freed up workers to innovate and explore new fields. As an example, at one time, up to 90 percent of the workforce was in agriculture; by the late nineteenth century, that number had dropped to 50 percent, and today, less than 5 percent of workers in Europe are employed in agriculture.

Today, many professions, like bank clerks, switchboard operators, or blacksmiths, have been replaced by new professions such as web designers, social media managers, and vloggers. Skills like long division, map reading, and manual typesetting have become obsolete. However, this process of creative destruction is not without its victims. Regions once reliant on heavy industries, such as the Ruhr area in Germany, the Rust Belt in the United States, and Northeast England, now face economic challenges as their industries decline. Even though overall wealth grows and productivity increases, change does not affect all equally and some individuals suffer much more than others even if, in the end, society as a whole is better off. The impact of AI and digitization on welfare, particularly among younger generations, is multifaceted. While AI and automation boost economic growth in some regions, rising concerns about inequality and job loss are driving public anxiety and distrust in institutions, particularly fears of unemployment due to large-scale automation of cognitive tasks.[30]

Nevertheless, predictions on the effect of AI and automation on labor and welfare are very divergent. Take, for instance, studies done by the leading research and consulting firm Gartner. Their 2014 prediction that by the mid 2020s at least a third of the jobs globally would be at risk of automation,[31] flipped radically in 2018 to a claim that millions of new jobs would be created after 2020,[32] to land on their current, much milder claim that *"Generative AI will impact many workers, but the timing and degree of impact will vary."*[33] But, though it may be tempting to dismiss these fluctuating predictions as speculative, history teaches us that technological revolutions always bring disruption. Without proactive steps, the transition to an AI-driven economy will have costs, and many communities may suffer.

Currently, major tech companies like Google, Amazon, and Microsoft hold immense power through their control of critical resources such as data, hardware, and computational infrastructure. This concentration of power enables these corporations to dominate the direction of AI development, shaping its future according to their priorities. Smaller companies, independent researchers, and even governments are often left at a structural disadvantage, unable to compete with the vast resources of these tech giants. This imbalance raises serious concerns about equitable access to the benefits of AI and the democratic distribution of power. If we are to create a future where AI serves society as a whole—rather than consolidating power in the hands of a few—it is essential that access to AI tools, knowledge, and opportunities be democratized. Equitable access to the power of AI means ensuring that people from all backgrounds, industries, and regions have the opportunity to contribute to and benefit from AI advances. This cannot happen without a concerted effort to break down the barriers to participation, especially those posed by the dominance of large tech companies.

Education plays a crucial role in leveling the playing field. To truly democratize AI, we must invest in widespread education initiatives that focus not only on digital literacy, data science, and AI-related knowledge in schools, universities, and workplaces but also integrate humanities and social sciences into these educational efforts. Though technical skills are crucial, understanding the broader societal, ethical, and philosophical implications of AI is equally important. Education in areas like ethics, philosophy, sociology, and psychology helps individuals critically engage with the consequences of AI on human behavior, social structures, and individual rights. Lifelong learning programs must be prioritized to equip workers not only with the technical skills needed to thrive in an AI-driven economy but also with the soft skills—like communication, critical thinking, empathy, and collaboration—that are essential in navigating the human aspects of technological change. These skills enable individuals to make informed decisions about how

AI is developed and applied in different sectors, ensuring that it remains aligned with human values and serves the broader interests of society. A well-rounded approach to education that balances STEM disciplines with social and humanities perspectives will empower individuals to navigate the complex interplay between technology and society. It will prepare them to contribute not just as technical experts but also as ethical leaders and responsible citizens in an AI-driven world. Only by fostering both technical competence and social awareness can we ensure that AI's benefits are shared equitably and used to improve lives in a way that is just, transparent, and inclusive.

Moreover, policymakers have a responsibility to ensure that the development and deployment of AI technologies are guided by principles of fairness, transparency, and accountability. This means stepping up the power of policymakers to engage at an equal level with the industry leaders, and involving diverse stakeholders in the decision-making process—workers, communities, and smaller businesses must have a voice in shaping AI's future. Ultimately, achieving equitable access to AI requires a balance of power among tech giants, governments, and the public. By investing in education and ensuring democratic oversight of AI development, we can build a future where AI serves the interests of all, not just the privileged few.

Key Takeaways and Reflections

The power paradox reveals the tension between the growing capabilities of AI and the diminishing control democratic institutions have over its development and application. As AI systems become more powerful, the ability to direct their impact and ensure their alignment with societal values becomes increasingly challenging. At the same time, the concentration of

AI development and resources in the hands of a few large corporations raises concerns about accountability, transparency, and equitable access. These organizations wield significant influence over how AI is designed and deployed, often prioritizing profit and control over the broader societal good. This imbalance reinforces existing power structures and exacerbates inequalities.

AI has immense potential to address global challenges, but realizing this promise requires a fair distribution of its power and a commitment to shared human values. Achieving this demands a collective effort: public participation, transparent decision-making, and inclusive governance are essential to ensure AI serves everyone, not just a select few. Strengthening regulations and fostering international cooperation can further align AI development with ethical standards and societal needs, ensuring it contributes meaningfully to the common good.

7

The Superintelligence Paradox

SUPER AI OR COLLECTIVE INTELLIGENCE?

A BOOK on AI paradoxes would be incomplete without addressing superintelligence. Given the challenges in defining intelligence, its computational aspects, and its ethical and societal implications, the concept of superintelligence adds another layer of complexity. To begin this discussion, we must first clarify what superintelligence is and how it differs from artificial general intelligence (AGI). AGI refers to AI systems capable of learning, understanding, and performing any intellectual task a human can, adapting across different domains. Essentially, AGI aims to create machines that think and reason like humans. It is often seen as a stepping stone to superintelligence, defined as AI that surpasses human intelligence in all aspects, including problem-solving, creativity, and decision-making. In contrast, today's AI systems are referred to as "narrow AI," in that they are able only to perform a very limited set of skills and in specific contexts. The key differences between AGI and superintelligence are their scope and impact: even if both represent major milestones in AI development, AGI marks the transition to machines that think like humans and superintelligence rede-fines the limits of cognition itself.

A key issue here is the very nature of AGI and superintel-ligence. Although they are typically viewed as computational systems focused on cognitive abilities, machine intelligence is very different from human intelligence, as we have seen in chapter 3. Moreover, human-made social systems—such as teams, organizations, and societies—also exhibit intelligence, which should be a trigger to rethink the way we conceive arti-ficial intelligence. Social systems, much like AI, evolve through decisions, actions, and innovation. Their ability to adapt, solve complex problems, and sustain large-scale operations mirrors the traits often attributed to superintelligent AI. In fact, AI itself is a product of human collaboration. Recognizing social structures as forms of intelligence reshapes how we think about AGI and superintelligence.

By shifting our perspective, we can explore a broader and more meaningful understanding of AGI and superintelligence, one that highlights the power of collective intelligence and promotes responsible AI development. Human teams and soci-eties, like AI, are designed to enhance our capabilities. They do more than organize people; they create intelligence greater than that of any individual, enabling complex problem-solving, innovation, and achievements beyond what one person could accomplish alone. In this sense, they function as a kind of arti-ficial superintelligence, emerging through collaboration and shared knowledge. And this is the essence of the superintelli-gence paradox:

The Superintelligence Paradox
The more we chase AGI, the more we discover that true superintelligence lies in human cooperation.

From this perspective, the true value of AI progress lies in leveraging its potential to enhance our collective intelligence. More intelligent AI could serve as a powerful tool, assistant, or partner, amplifying human teams and societies by optimizing decision-making, problem-solving, and coordination beyond human limits. Integrating AI into social structures could help manage complexity at an unprecedented scale—processing vast data, predicting outcomes, and suggesting strategies that exceed human cognitive capacity.

Here, it is important to revisit the differences between human and artificial intelligence, as discussed in chapter 3. Unlike AI, human intelligence is deeply tied to survival, adaptation, and the pursuit of meaningful lives. As social beings, we constantly balance individual needs with the collective good, adapting our actions based on context, available tools, and our physical presence in the world. In contrast, advances in AI, and particularly AGI, focus mostly on cognitive processing, independent of real-world embodiment or intrinsic motivation. This creates a critical gap: AGI lacks an inherent understanding of its limits and of when enough is enough, as well as the ability to balance self-preservation with societal needs or define a purpose for its own reasoning. But, just as calculators perform complex computations without true intelligence, AI systems can be highly useful without possessing genuine understanding.

True intelligence goes beyond performing complex cognitive tasks; it also requires contextual awareness, adaptability, and the ability to navigate competing priorities. Moreover, human intelligence needs a biological body. Although AGI is designed to generalize across domains and solve complex problems, it lacks intrinsic understanding, self-awareness, or independent purpose. Its decision-making remains bound by programmed algorithms and predefined instructions, which dictate how it processes external feedback, learns patterns, and applies reinforcement mechanisms. As it optimizes for new goals and situations, adaptability is ultimately a function of its underlying design. Current narratives often anthropomorphize AGI, depicting it as driven by survival instincts: seeking to preserve goals, establish principles, or ensure self-sufficiency. However, this framing is misleading; AGI does not possess inherent desires or a will to survive.[1] Instead, such narratives serve to justify control by those developing and deploying AI, reinforcing the idea that AGI must be contained or aligned for safety. In reality, AGI is a product of human design, shaped by programmed instructions and optimization processes. Rather than assuming it will default to rigid goal preservation, we can explore alternative models that prioritize adaptability, where AGI can modify or even discard its objectives when they become obsolete, ensuring it remains flexible and aligned with evolving human needs.

If we recognize collective intelligence as a form of super-intelligence, then the goal of AGI and computational super-intelligence should not be to replace human intelligence but to enhance it. Rather than striving to create monolithic, isolated, all-powerful systems, the emphasis should be on how AI can amplify our collective ability to think, collaborate, and solve problems. In this view, AGI and superintelligence should be developed as tools that facilitate communication and coordination across large groups, aligning individual efforts toward shared goals. By identifying patterns and solutions that might otherwise go unnoticed, AI could provide valuable insights that drive innovation and social progress. Moreover, by augmenting human capabilities, AI-driven collective intelligence could play a crucial role in addressing global challenges that require large-scale cooperation, such as climate change, pandemics, and sustainable resource management. Unfortunately, this perspective is often overlooked in current AGI and superintelligence research. The prevailing focus is on creating autonomous systems that are all-knowing and independent—designed to act on their own—rather than systems that work alongside humans to extend our capabilities and enhance collective intelligence, that is, autonomous systems that work with us rather than instead of us.

My main concern is that the current pursuit of superintelligence may be more of a distraction than a solution to today's pressing challenges. The idea of creating an "artificial universal problem solver" suggests not only a misunderstanding of AI but also a deeper failure to grasp the nature of our problems. As the cryptographer Bruce Schneier once observed,[2] "*If you think technology can solve your security problems, then you don't understand the problems and you don't understand the technology.*" The same logic applies here.

More than just a distraction, this approach to superintelligence is potentially harmful. It diverts focus and resources from more immediate, actionable solutions while concentrating power in the hands of those controlling AI development, as discussed in chapter 6. This reminds me of a childhood story. Whenever our living room became a minefield of LEGO blocks, my mother would call on us to clean up. While we grumbled and sorted pieces, one sibling always went for a different approach: rather than helping, he would start building a *"LEGO pick-up machine"* out of LEGOs, a grand idea that only created more chaos, inevitably sparking another sibling quarrel. I couldn't help but think of this when the former Google CEO Eric Schmidt argued that developing AI should take precedence over climate change mitigation. He claimed that climate goals are unattainable due to a lack of global organization and that AI's energy consumption should not slow its progress. Instead, he suggested that AI itself might eventually solve environmental challenges, stating: *"I'd rather bet on AI solving the problem than constraining it and having the problem."*[3] This mindset, that AGI will be the ultimate solution to all our problems, feels like the adult equivalent of the LEGO pick-up machine: an impressive display of technological ambition that ultimately misses the point. Just as my brother hoped for an effortless way to clean the room, we risk postponing urgent climate action in the belief that a future AGI will miraculously fix what we already have the tools to address. The real challenge lies not in developing an all-powerful AI but in making the necessary structural, social, and political changes to combat climate change today.[4]

AGI Is Not Inevitable

As discussed in chapter 3, AI is human-designed with specific objectives. Its development path, including the role of AGI and superintelligence, is a choice, not an inevitability, despite discussions suggesting that AGI is imminent. The trajectory of AI is always a reflection of the choices we make.

To qualify as AGI, a system must mirror human intelligence by learning, reasoning, and solving problems across various contexts. It should process sensory data, understand language, and interact meaningfully. Some definitions also include social intelligence, and the ability to navigate human interactions and respond to emotions appropriately. The recent advances in large language models (LLMs) have been hailed as stepping stones toward AGI, given their impressive performance in language processing and problem-solving. However, despite their versatility, significant limitations prevent LLMs alone from achieving true AGI. As many experts have pointed out,[5] LLMs lack true understanding; they operate by predicting statistically probable responses rather than engaging in meaningful cognition. Their behavior is not the result of reasoning or intentionality but of pattern recognition at scale.

Moreover, LLMs struggle with generalization across domains. Although they excel in tasks closely aligned with their training data, they falter when faced with novel situations that require flexible, cross-domain reasoning, one of the fundamental hallmarks of AGI.[6] Their reliance on massive amounts of curated data further limits their adaptability and increases the potential for misuse and harmful consequences.[7] In contrast, a truly general intelligence would need to learn efficiently from minimal information and apply knowledge in fundamentally new ways.

Having said this, it is important to recognize the risks of anthropomorphizing LLMs. Although it is natural to use humanlike language to describe technology (e.g., "my computer hates me" or "my car didn't want to start"), doing so in reference to LLMs becomes particularly misleading. As the philosopher and cognitive scientist Daniel Dennett describes, this tendency, known as the intentional stance, is a strategy for interpreting the behavior of an entity by treating it as if it were a rational agent.[8] However, LLMs are simultaneously so very different from humans in their construction yet often strikingly humanlike in their behavior, "that we need to pay careful attention to how they work before we speak of them in language suggestive of human capabilities and patterns of behaviour," as Murray Shanahan, a senior scientist at DeepMind, warns.[9] At the most fundamental level, he continues, LLMs have no communicative intent, no awareness of whom they are interacting with, and no understanding of the context or purpose of a conversation. Their nature is fundamentally unlike our own, presenting an uncanny mixture of less-than-human and super-human capabilities—sometimes eerily humanlike, other times entirely alien. In a sense, as I often describe, LLMs are akin to a "cognitive Frankenstein's monster" stitched together from vast amounts of data and complex algorithms, piecing together fragments of human language in a way that appears intelligent. However, like Frankenstein's monster, they lack genuine under-standing and intentionality. They do not "think" or "know"; they merely simulate patterns extracted from their training data. This fragmented, imitative nature should serve as a cautionary reminder against anthropomorphizing AI systems—a warning that, although they may convincingly mimic intelligence, they do not possess the cognitive depth, self-awareness, or intrinsic motivation that human intelligence entails.

The assumption that AGI is an inevitable outcome of AI progress is deeply flawed. As discussed earlier, AI is not a self-propelling force of nature—it is a human-made artifact, shaped by deliberate design choices, research approaches, and the values that guide its development. Even though much of the current discussion implies that AGI is just around the corner, this assumption is misleading. The question is not just *when* AGI will happen, but *whether* it will happen at all and, more important, *whether we actually want it to happen.*

A key argument against the inevitability of AGI is its com-putational intractability.[10] Recent research suggests that even under idealized conditions, the sheer computational resources required to transition from narrow AI to AGI with current tech-nological constraints, are prohibitively large, making the goal of AGI not just distant, but potentially unfeasible.[11] The complex-ity of tasks increases exponentially with intelligence, meaning that scaling current AI approaches may never be sufficient to achieve true general intelligence. Beyond technical feasibility, the notion that AGI is an inevitable outcome of AI progress is itself flawed.

Additionally, AGI, like AI itself, is not merely a technical challenge; it is a sociopolitical construct. Its development depends on the priorities of those funding and directing AI research, whether governments, corporations, or academic institutions. Many claims that AGI is inevitable serve as a rhetorical tool to justify its pursuit, often to secure funding, influence policy, or consolidate power. As a result, the dominant narrative frames AGI as something that will happen, rather than something that is actively being shaped by decisions made today.

The way AGI is conceptualized is a direct reflection of the methods and values embedded in its design. Whether we choose to pursue large-scale deep learning models, symbolic AI, or hybrid approaches will dictate the limitations and capabilities of what we call AGI. Some argue that bigger models trained on larger datasets will eventually "crack" intelligence, while others believe that alternative approaches, such as neuromorphic computing, hybrid AI, or evolutionary algorithms, may be necessary to move beyond the current paradigm.

Furthermore, the alignment of AGI with human values is not a purely technical issue but also a sociopolitical one. Whether AGI is designed to optimize economic efficiency, enhance human well-being, or serve military and surveillance interests will profoundly shape not only how it is built but also, more important, its impact on the world. The question is not simply, can we build AGI, but what kind of AGI are we trying to build, and for whom?

One of the most overlooked aspects of the AGI debate is that intelligence is not a single, monolithic property, but rather a diverse and distributed phenomenon. As discussed earlier, human intelligence is inherently social, developed and sustained through collaboration, culture, and shared knowledge.[12] AGI is often imagined as an isolated system, an omniscient machine capable of independently solving all problems, but this vision ignores the reality that intelligence, in its purest form, has always been collective. Historically, human civilizations have created artificial systems that enhance and extend our collective intelligence, from writing systems and legal frameworks to scientific institutions and digital networks. If we redefine AGI not as a singular entity but as a continuation of this tradition of artificial systems designed to augment rather than replace human intelligence, then AGI is something we have been building for millennia.

This shift in perspective challenges the techno-deterministic narrative that AGI is an unavoidable outcome of AI progress. Instead, it reveals that AGI is not an external force that "happens" to us, but a reflection of our collective choices. The question is not whether AGI will emerge but how we choose to define, design, and deploy AI systems that shape our future. The myth of AGI's inevitability serves those who wish to portray it as an unstoppable force, deflecting responsibility for the decisions shaping its trajectory. But AGI is not a law of nature; it is a human construct. It will not emerge accidentally, nor will it develop in a vacuum. If AGI ever comes into existence, it will be the result of intentional design, political agendas, and societal values. Rather than asking when AGI will arrive, we should be asking:

- Who benefits from AGI, and who bears its risks?
- What alternative paths exist for AI development?
- Should AGI be pursued at all, or should AI be designed to enhance collective intelligence instead?

The answers to these questions will determine whether AGI and superintelligence remain a speculative fantasy or become a deliberate and meaningful step toward enhancing human potential.

When Will Airplanes Lay Eggs?

In current discussions, AGI is generally seen as the first step toward achieving superintelligent AI. However, when, or even if, we will reach AGI remains a matter both of debate and of personal interpretation. If we define superintelligence as collective intelligence, such as a human team or society, as I do at the beginning of this chapter, then superintelligence already exists and has always existed. Yet, this perspective is not what drives today's technological efforts. Instead, superintelligent AI is often envisioned as a godlike entity: an all-knowing, all-powerful system that can autonomously solve problems, make decisions beyond human reach, and even reshape the world.

This idea, reminiscent of science fiction, is surprisingly common even among leading AI researchers. Many assume that superintelligence will take the form of a single system with independent thought, emotions, and deep understanding, a sentient entity capable of solving all problems, controlling the world, or even destroying it. However, given that AI development is not about crafting fictional narratives but engineering real-world systems, this assumption deserves deeper scrutiny.

For example, Yoshua Bengio and colleagues describe frontier AI systems as the most advanced AI models under development: highly autonomous, with the potential to surpass human control in critical domains.[13] These systems are envisioned as possessing immense computational power and the ability to act independently, possibly pursuing goals misaligned with human intentions. Once they reach a certain threshold of capability, they could engage in strategic reasoning, manipulation, or hacking, making them increasingly difficult to regulate or contain. Their paper warns of risks such as irreversible loss of control, large-scale social harm, and threats to human safety and autonomy. Notably, these descriptions suggest a vision of AI that borders on sentience: machines that could self-replicate, evade human intervention, and accumulate influence. This portrayal fuels concerns about unchecked power, reinforcing the godlike narrative. But to fully assess these risks, we must critically examine a key underlying assumption: Does intelligence necessarily imply sentience or consciousness? If not, then the fear of AI developing independent desires or self-preservation instincts may be fundamentally misplaced.

For me, the idea that a single AI system could decide to destroy the world, or single-handedly solve all our problems, seems not only oversimplistic but mostly unrealistic and a bit naive. It reflects a misunderstanding of the complexities of real-world issues[14] and, more worrying, it suggests an attitude of shifting responsibility onto the technology and, in the process, absolving oneself of any accountability: the system knows all, so the blame for whatever happens can be given to the system. More than the risks of the technology, it is this mindset that creates an utterly dangerous situation where outcomes are blamed on the machine, akin to an all-too-convenient *"the computer says no"* excuse. Such an attitude overlooks the essential role of human accountability in the use and outcomes of AI technologies. Unfortunately, as we saw in chapter 6, there are powerful forces that have everything to gain by propagating this idea.

Moreover, the current approach to AGI and superintelligence leaves aside many important questions, starting with the questions, *Why* would an AI system want to control or destroy the world? and *How* will such a system arise from the computational, statistical models that are the current level of AI development? First, as we have discussed in depth in this book, human intelligence is multifaceted and much more than just the cognitive abilities needed to use language, reason, or solve problems. In this sense, the idea of machines approximating or surpassing human intelligence, in all its complexity, can be compared to the idea that airplanes will soon be laying eggs, just because we keep improving their flying capabilities! This comparison highlights the absurdity of expecting a machine—a nonliving, mechanical artifact— to attain the full spectrum of human intelligence, which is deeply entwined with our physical bodies, consciousness, and lived experiences. Just as an airplane, no matter how advanced, will never perform a biological function like laying eggs, AI systems, no matter how sophisticated, are unlikely to develop consciousness or replicate the full breadth of human intelligence.

Risks and expectations about superintelligence stem both from a misunderstanding of the fundamental nature of machines and from mixing up different meanings of consciousness. Consciousness can refer either to **awareness** of oneself and the environment or to the **subjective experience** of what it is like to perceive and feel things. Awareness involves the ability to recognize and respond to surroundings and internal states—that is, perceiving what is happening around you and within you. For example, when you see a tree, you are aware of its presence and can describe it. On the other hand, subjective experience is about the personal, inner feeling of those perceptions—that is, what is it like for you, specifically, to see the tree, including the unique way you perceive its color and shape, and the emotions it might evoke in you. Awareness is more about external recognition and understanding, while subjective experience is about the internal, personal quality of those experiences. Machines can be built to be "conscious," in the sense of awareness, but it is a big stretch to expect that machines, designed to execute tasks within a specific, defined scope, will evolve into beings capable of subjective feeling. Of course, this does not mean that AI is not useful or that it cannot support us to address complex cognitive problems. In the same way that airplanes are useful, but at the same time, their use is an increasing liability to the environment and therefore to our lives, the capabilities of AI systems also come with constraints and problematic consequences.

Is AI the Superintelligence We Need?

Discussions about AGI, superintelligence, and existential risks dominate much of the contemporary AI debate. This discourse has split the field into different camps: those who believe superintelligence is inevitable and those who do not; those who fear AI-driven human extinction and those who focus on more immediate, tangible risks. At the beginning of this chapter, I made it clear that superintelligence already exists, but not in the form most AI researchers imagine. Rather than an all-knowing AI system, superintelligence has always been collective, emerging from human cooperation. What I do not believe in is the narrow, techno-utopian vision of an omniscient AI capable of solving all problems autonomously. Although AI advances bring significant opportunities and risks, discussions must remain grounded in reality rather than science fiction.

The assumption that AGI will be desirable is rarely questioned. Simply put, if AGI and superintelligence emerge and could potentially replace humans in some domains, should we allow that to happen? The dominant narrative assumes that creating a system that surpasses human intelligence is a natural progression of AI development. But what exactly is the problem AGI is meant to solve?

AI has already demonstrated remarkable success in specialized areas—detecting diseases in medical scans, optimizing logistics, and generating creative content. Yet, none of these successes require an all-powerful AGI. Most of humanity's pressing challenges, from climate change to economic inequality to political instability, are not the result of insufficient intelligence. Rather, they stem from systemic and structural issues that an autonomous AI system would be ill-equipped to address. Instead of addressing the difficult but necessary social and political changes, some see a future AGI as the solution to these problems. As we saw earlier, the former Google CEO Eric Schmidt exemplifies this mindset: when discussing climate change, Schmidt argued that efforts to mitigate emissions are futile without AI. He suggested that rather than constraining AI development due to its energy consumption, we should *"bet on AI solving the problem"* itself. This line of thinking assumes that AI will somehow compensate for political inertia and systemic failures.

By focusing on AGI as a distant, all-encompassing solution, we risk neglecting the more immediate and practical ways AI could be used to enhance human decision-making today.

If we abandon the idea that AI must function as an autonomous superintelligence, we can explore alternative ways to design AI that enhances human intelligence rather than replacing it. That AGI will outperform humans in some cognitive functions is nothing new; calculators have done so for decades. However, rather than devaluing human intelligence, this shift highlights uniquely human capabilities. Philosophers have long warned of scenarios where humans become mere maintainers of machines rather than active participants in society.[15]

Measuring AGI against human intelligence assumes a singular notion of intelligence, yet intelligence varies across individuals, cultures, and contexts. By designing AGI to surpass specific benchmarks, developers implicitly select particular models of intelligence—often mirroring their own biases. In 2014, the chatbot Eugene Goostman was claimed to have passed the Turing test, but this was largely due to its persona as a thirteen-year-old Ukrainian boy,[16] which masked inconsistencies. This suggests that such tests reveal more about human perception than machine intelligence. Moreover, intelligence tests and benchmarks used in AI often reflect Western, industrialized perspectives, prioritizing cognitive skills like logic over social intelligence or cultural knowledge. This bias risks reinforcing societal inequalities by favoring intelligence types associated with privileged groups. If AGI is designed to excel in areas dominated by these groups, it could exacerbate social divides.

The belief that AGI will solve all our problems is also misguided. It distracts from the need for human collaboration and shifts power to those who control AI systems. Narratives about superintelligence often include scenarios like Nick Bostrom's "paperclip maximizer,"[17] where an AI single-mindedly pursues a goal at the expense of humanity. These examples illustrate the importance of aligning AI with human values across generations. Visions of superintelligence frequently depict machines solving complex issues like climate change through vast data analysis. Enthusiasts claim that AI could reduce CO_2 to pre-industrial levels or cure all diseases. However, such optimism overlooks critical flaws: AI could propose harmful solutions, such as turning oceans into carbonic acid as a way to clear carbon from the air, or exacerbating health inequities through biased medicine, as happened when a widely used AI algorithm in U.S. hospitals was found to systematically discriminate against Black patients by underestimating their health risks, leading to less access to critical medical care.[18] Existential risks from AI are not just about future domination but also about the current societal structures that enable its unchecked deployment. Another unhelpful alternative is that AI will provide simple but undesirable solutions, such as proposing a ban on all fossil-fuel vehicles, from cars to airplanes. This would do the trick but ignores larger societal needs, and it would require a level of political will that no current government would dare to attempt, most specifically not a decision that either the U.S. or the Chinese government would take even if most of the vehicle emissions originate from these two countries.[19]

Instead of asking how to prevent AGI from harming us, we should question why we would delegate such power in the first place. Systems that mindlessly optimize or pursue goals without reassessing their consequences—like the infamous paperclip maximizer—should not be mistaken for superintelligence. A truly superintelligent system would recognize when to stop, evaluate its impact, and seek guidance rather than blindly following objectives to disastrous ends. The real danger lies in assuming that AI must operate autonomously without accountability.

The notion of superintelligence as an omnipotent entity is deeply rooted in philosophical traditions. From Plato's philosopher-kings to Nietzsche's Übermensch, thinkers have long speculated about the pursuit of ultimate knowledge and control. This ambition finds a modern parallel in transhumanism, a dystopian movement championed by several Big

Tech elites who believe that human enhancement through AI, genetic engineering, and cybernetic augmentation is an inevitable and desirable future. However, transhumanism is not about human progress; it is about control. It risks deepening inequality, eroding autonomy, and reducing human life to a mere technological optimization problem. Transhumanism is the modern Promethean hubris, an unchecked ambition that, much like the myths of Frankenstein, the golem, and the sorcerer's apprentice, warns us of the dangers of creating something beyond our control. The belief in an AI-driven post-human future reflects a blind faith in technology over humanity, prioritizing artificial intelligence over ethical reasoning, community, and wisdom. A truly advanced intelligence would seek not to dominate but rather to understand its own limitations, collaborate, and defer to human values. Rather than placing our hopes in transhumanist fantasies, we should focus on strengthening our human capabilities—through ethics, wisdom, and collective intelligence—not surrendering them to soulless algorithms. AI should be developed as a tool for humanity, not as an autonomous force that dictates the fate of civilization.

Despite AI's rapid progress, it remains far from humanlike intelligence. Common sense, contextual understanding, and humanlike movement are still significant hurdles. While AI continues to evolve, our focus should remain on ensuring it serves humanity rather than defining intelligence by artificial standards.

The Real Risk and How to Proceed

One of the main arguments for the existential threat posed by AI is that when AI systems become intelligent enough, they may hurt humanity. Not long ago, Turing Award winners Yoshua Bengio, Geoffrey Hinton, and Andrew Yao, along with several other experts, warned that *"Without sufficient caution, we may irreversibly lose control of autonomous AI systems, rendering human intervention ineffective,"* and pointed to the need to *"prepare for the largest risks well before they materialize."*[20] Scary, to say the least. But what is really at stake in these warnings? Should we be afraid of losing control over AI? I fear that we should be more concerned about losing control over those who develop, own, and deploy these systems.

The issue of "existential risk" mostly means that we need to revisit the notion of intelligence. If such systems are truly intelligent, they are intelligent enough to understand the context in which they follow their goals. But, as they are usually described, those "superintelligent" systems are self-absorbed and oblivious of their social context and, thus, very far from human intelligence. In fact, as the computer scientist and Turing Award winner Yann LeCun has pointed out, it is not intelligence we should be concerned about but abuses of power; we should be more concerned about controlling those who control AI than about AI as a technology. I have been a research leader for many years in autonomous agents and multiagent systems. In this field, we have transferred many of the societal solutions to regulate, monitor, and control complex entities to socio-technical equivalents. Techniques such as contracts, institutions, norms, and regulations have a role in computational systems but cannot be fully operationalized without a human, and organizational, grounding. My concern is that a narrative of "we don't know how to control AI; it is going to destroy us" only helps to create panic and gears us further from applying and developing what we do know. Even though algorithmic and mathematical approaches will be needed, most importantly we need to apply societal means to ensure alignment and governance, not only technological solutions.

The current focus on existential risks distracts from the real, present-day dangers posed by AI systems. Going forward, the focus needs to be on understanding and addressing the actual harms caused by AI today.

The most relevant questions about AGI are those that no one is asking: Is it desirable? Is it inevitable? Is it necessarily computational? Perhaps the most desirable version of AGI is not replacing but augmenting human intelligence: empowering humans to tackle challenges while preserving our capacity for moral, cultural, and emotional decision-making. The measure of AGI should not be how much it replaces us, but how much it enhances what we value most. Inevitability is a myth we tell ourselves: progress in AI will march forward regardless of human intent. It brings with it the idea that "AI happens to us," which we have discussed earlier in this book. But AGI is not a force of nature: it is a mirror of our collective will and responsibility.

AGI is something we've been building for millennia: artificial social systems that amplify our collective capability. True AGI isn't a computational, alien, construct; it's the latest chapter in humanity's story of creating tools to enhance our shared intelligence. Its utility is not in prescribing the "optimal" decision but in supporting and amplifying the diversity of human reasoning. The main issue is how we will choose to define and shape it. Will it be a tool for collaboration or a mechanism to allow a few powerful to control the rest of us? A way to enhance human complexity or to reduce us to machine-readable data?

The goal is to create AI systems that complement human cognitive strengths and address weaknesses, ultimately enhancing our collective problem-solving capabilities. If we make AI systems intuitive and accessible, they can serve a broad range of users, making these tools integral to improving collaboration across diverse groups. Recent work on human-centered AI approaches highlights the role of AI in augmenting collective intelligence, particularly in enhancing communication and coordination within and between human groups.[21] Using a hybrid approach, AI can facilitate real-time collaboration by seamlessly integrating human expertise with machine intelligence, enhance data sharing by combining automated processing with human contextual understanding, and support decision-making processes that are more aligned, efficient, and ethically informed. Leveraging AI's pattern recognition capabilities alongside human intuition, this hybrid approach can generate deeper insights, drive innovation, inform policymaking, and tackle large-scale challenges that surpass the capabilities of either humans or AI alone.

Key Takeaways and Reflections

The superintelligence paradox highlights the tension between humanity's aspirations for advanced AI and the challenges of ensuring its alignment with human values. Although the concept of superintelligence (AI surpassing human intellect in all areas) captures the imagination, it also raises critical concerns about governance, control, and societal impact.

True superintelligence does not reside in AI itself but in humanity's collective intelligence, our ability to collaborate, govern responsibly, and align technology with ethical principles. Instead of delegating control to machines, the focus should remain on fostering human participation, reflection, and decision-making to address the complex societal problems that AI alone cannot solve. The superintelligence paradox reminds us that AI's future will not be defined solely by technological breakthroughs but also by our choices and actions today. It is a call to prioritize shared human values, inclusive governance, and global cooperation to ensure that AI remains a tool for collective progress and well-being.

8

The Solution Paradox

THE TECHNOLOGY TRAP

THIS BOOK may have felt a little like a roller-coaster, depicting AI as both a solution and a challenge—simultaneously beneficial and harmful. AI needs regulation while fostering innovation, it has the potential to create a more equitable world but it is not without risks, and it is widening the gap between those who reap its benefits and those who bear its consequences. Are the costs higher than the benefits? How do we even begin to measure them? So many directions, so many questions. In many ways, AI is no different from cars. Cars allow us to travel greater distances, connect with loved ones, and live farther from work. Yet, they also cause accidents that claim lives and can even be misused, as tools to cause harm. Like AI, car technology is evolving every day and requires continuous regulation. Cars today are much safer and more efficient than those from fifty years ago, and they will continue to evolve. But the core principle of car regulation remains the same: ensuring the safety of people and the environment. In the same way, AI can provide tremendous benefits to society when developed and used responsibly. However, just as we regulate cars to protect ourselves from accidents and misuse, AI also requires safeguards to prevent harm and ensure it serves humanity's best interests. The challenge ahead lies in finding that balance by encouraging innovation while protecting society from the unintended risks of this powerful technology. In this final chapter, we explore the paradox of AI technology: AI is a solution that can offer immediate benefits yet frequently overlooks broader societal impacts.

The Solution Paradox
Solving problems with technology often creates more problems.

When we listen to the mainstream AI narrative, the widespread adoption of AI seems to be taken as a given, an inevitable development, based on the implicit belief that AI technologies can solve anything if only we take care of the potential side effects. This idea is called **techno-solutionism**: the belief that social, political, and environmental problems can all be solved with the right technology. Techno-solutionism is often characterized by an unwavering optimism about the potential of technology to improve human lives and address global challenges. The appeal is clear: technological solutions often appear faster and less complicated than pursuing societal or systemic changes. However, there are several problems with this approach, starting with the fact that technological fixes often treat symptoms rather than addressing the causes. Moreover, a techno-solutionistic mindset undervalues the importance of human judgment, cultural context, ethical considerations, and the complex, nuanced understanding that people bring to decision-making processes. Although technology can offer powerful tools, relying solely on it with-out considering the broader human and societal implications can lead to oversimplified solutions that don't fully address the underlying issues. Thinking that technology can solve com-plex problems leads to oversimplification of those problems and, consequently, to the creation of new problems. Techno-logical considerations alone cannot dictate the conditions for social life, but rather need to reflect how we want our lives, our societies, and our technology to be.

As can be seen from the earlier chapters, AI is not magic, nor is it the solution to all of our problems. In this last chapter, I challenge both the view that AI, if only constructed in the right way, can be a catch-all solution to a range of social and envi-ronmental problems, as well as the idea that AI will ultimately take over the world and destroy us all. Many experts, including myself, argue that we need a responsible, ethical, and trustwor-thy approach to AI development and use. At the same time, calls for stronger AI governance and concerns about the lack of ethics in its development and use are fast growing, with gov-ernments, corporations, and social organizations falling all over each other in the rush to publish yet more recommendations, principles, or guidelines. Such commitment to an accountable, responsible, transparent approach to AI, where human rights and societal values are the leading principles, is laudable and a much needed development, one to which I've dedicated much effort and research in the past few years. But is that enough? And what does such an approach truly entail?

Ensuring truly responsible development and use of AI requires more than principles and regulation. It starts with a clear understanding of what AI is, what it is not, what it can do, and what it cannot. A simple, clear narrative about AI is crucial to help everyone participate in discussions about the role of AI in society. However, as we have seen, some may prefer to keep things unclear and complicated. It is essential to tackle the challenges and misconceptions surrounding AI through education, clear communication, and collaborative policymaking. We also need insights from various disciplines, particularly those that blend technology with social sciences and humanities.

AI-Solutionism

AI-solutionism is a particular type of techno-solutionism, rooted in the belief that artificial intelligence can single-handedly resolve a wide range of societal challenges. This mindset is appealing because it offers seemingly straightforward, efficient solutions to complex issues. However, AI-solutionism is fundamentally flawed.[1] A solutionism approach often leads to the rushed deployment of technological solutions without a thorough understanding of the deeper underlying problems they aim to address.

For instance, complex issues like climate change, political polarization, and global inequalities are not just technical challenges; they are deeply intertwined with social, economic, and cultural factors. Climate change involves not only the science of reducing carbon emissions but also the politics of international agreements, the economics of energy transitions, and the social justice issues related to who bears the brunt of environmental degradation. Similarly, political polarization is not merely a problem of information dissemination that can be solved with better algorithms; it involves historical grievances, identity politics, and media dynamics that are far more nuanced. Global inequalities are rooted in a complex web of historical, economic, and social factors that simple technological solutions cannot unravel.

Society and technology are deeply interconnected, constantly shaping each other in a dynamic relationship. Social values, cultural norms, and political agendas significantly influence the development and use of technological innovations. In turn, these technologies reshape social practices, economic structures, and political dynamics, creating new opportunities and challenges. This mutual influence means that although technological advances can drive progress, they cannot alone determine social and political goals. Instead, these goals should be thoughtfully integrated into technological development to ensure that innovations align with and support broader societal values and ethical principles, ultimately promoting a more equitable and just society.

AI solutions often assume that problems can be clearly defined and that technology can offer exact answers. However, by focusing on efficiency and rationalism, AI often overlooks the complex and nuanced nature of real-world issues. In reality, these problems are multidimensional and require solutions that are not only technologically robust but also socially and ethically informed. Oversimplification can lead to unintended consequences, such as worsening existing inequalities, reinforcing biases, or creating new forms of exclusion. Therefore, while AI and other technologies can play a significant role in addressing societal challenges, they cannot do so in isolation. Effective solutions require a deep understanding of the issues at hand, considering the social, cultural, and political contexts in which they exist. Without this comprehensive approach, AI-solutionism risks creating more problems than it solves, ultimately failing to address the root causes of the challenges it seeks to overcome.

Moreover, AI systems do not operate in a vacuum; they are deeply embedded within social, economic, and political frameworks. As a result, the implications of AI deployment can be far-reaching, affecting various aspects of society in ways that are not always immediately apparent. This interconnectedness means that the impact of AI extends beyond the specific problems it is designed to address, influencing broader societal dynamics and potentially leading to unintended consequences. Take, for example, the COVID-19 pandemic, when AI was used to predict disease spread and to monitor social distancing. Although some of these applications were crucial, they also highlighted significant risks.[2] In fact, some AI models underestimated COVID-19's spread, particularly in regions with poor data, and overestimated the effectiveness of interventions like lockdowns, leading to overly optimistic predictions.[3] Additionally, AI-driven monitoring raised privacy concerns with potential long-term impacts on civil liberties, highlighting the limitations of AI in complex situations.

Similarly, AI-driven facial recognition is introduced in public services under the promise of enhanced security but risks exacerbating biases, particularly against marginalized communities. Research by the computer scientist and digital activist Joy Buolamwini and others has demonstrated that these systems can be much less accurate for certain groups, often resulting in wrongful identifications. In particular, Buolamwini's project Gender Shades specifically highlighted that facial recognition systems are particularly inaccurate for individuals with darker skin tones and for women. Her studies demonstrated that those higher error rates when identifying people of color can lead to wrongful identifications and further marginalization. This inaccuracy, combined with the potential for mass surveillance, raises serious privacy concerns and highlights the broader implications of deploying such technologies without sufficient oversight. These examples underscore the potential of AI but also the dangers of deploying it without a comprehensive under-standing of the broader context. AI can inadvertently worsen inequalities and amplify biases if not carefully designed and implemented. Therefore, a cautious approach, involving con-tinuous monitoring, ethical considerations, and public engage-ment, is essential to ensure AI technologies are used responsi-bly and in a way that promotes a more equitable society.

To move beyond AI-solutionism, a critical, multidisci-plinary perspective is necessary. Understanding social prob-lems should take precedence over seeking technological fixes. AI should not be viewed as a panacea; instead, its sociopolitical impacts must be thoroughly examined. Aligning technological advances with democratic and human rights principles is cru-cial for ensuring just and equitable outcomes. The rise of ethical guidelines and principles for AI highlights the increasing aware-ness of these issues. However, implementing and monitoring these principles is challenging. A balanced approach, combin-ing top-down and bottom-up strategies, is essential. Ethical frameworks must be adaptable to different cultural contexts, supporting a variety of interpretations and values. This flexi-bility ensures that ethical considerations remain relevant and evolve with societal changes. By integrating diverse perspec-tives and adapting to cultural contexts, AI development can better address ethical complexities and contribute to a more equitable and just society.

Addressing the limitations of AI requires a multidisciplinary approach that combines technological expertise with critical social theories. This approach is essential for tackling complex societal issues and ensuring that AI contributes to more just and equitable outcomes. By examining the sociopolitical impacts of AI, we can guide its development in ways that benefit society as a whole.

An AI Race to the Bottom?

One of the consequences of techno-solutionism is the "race to the bottom" effect, when businesses, organizations, or countries compete to lower standards, wages, regulations, or other factors to gain an advantage. This concept is often discussed in relation to the so-called AI race, which refers mostly to two different types of competition. First, the one going on between tech giants such as Microsoft, Google, or OpenAI. There is an intense competition to rapidly develop AI technologies, with the current focus being on the development of artificial general intelligence (AGI). These companies are striving to dominate the market, push technological boundaries, and secure leadership in AI innovation. NVIDIA, providing essential hardware, is a key player in this race. AI development is becoming increasingly commoditized, with some experts warning that large language models may soon be as cheap and abundant as rice, driving down quality. The rush to outpace rivals has also led to ethical oversights, with critics cautioning that reckless development threatens safety and stability. Labor exploitation is another concern, as companies rely on low-paid workers in digital sweatshops to train AI models. At the same time, excessive spending on AI, as seen in Big Tech, is squeezing profit margins and leading to unsustainable business practices.

Second, on the geopolitical front, the "AI race" is a high-stakes contest between global powers, mainly the United States and China. This competition centers on national security, economic supremacy, and control over critical technologies like semiconductors. For instance, the United States has implemented strict controls on semiconductor exports to China to limit its AI progress, leading to a "chip war." In response, China is strengthening its semiconductor industry and seeking alternatives to maintain its technological momentum. The global competition in AI intensified in early 2025 with the emergence of new players and shifts in international collaborations. Notably, China's DeepSeek[4] has developed the R1 AI model, which rivals leading U.S. technologies like ChatGPT but was created with reportedly significantly fewer resources. This development can disrupt the AI landscape, challenging the notion of U.S. dominance in the field.[5] At the same time, China's rapid AI development, particularly in surveillance and military applications, raises global concerns about security, ethical governance, and the potential shift in technological dominance from the West to the East. Beyond the United States and China, other nations and international organizations are actively engaging in AI development and governance. For instance, the Global Partnership on Artificial Intelligence (GPAI[6]), established in 2020, brings together countries like Canada, France, and Japan to promote responsible AI development, in close collaboration with OECD. Similarly, the United Nations has convened a High-Level Advisory Body on AI[7] to analyze global AI governance and foster international cooperation.

But are these races? And how much should we be concerned about the progress of such races? In athletics, a race is a speed competition, where competitors strive to reach a finish line on a predefined course as quickly as possible. But, for most amateur (long-distance) runners, the goal isn't to win the race but to push one's own limits. The true appeal lies in the satisfaction of reaching the finish line, ideally enjoying the process while doing it, as so well described by the author Haruki Murakami in the book *What I Talk about When I Talk about Running*. As a marathon runner, I am very familiar with the nature of races and can definitely say that AI is not such a race: there is no finish line, and current advances like generative AI or foundational models are merely the start of what's to come. Progress in AI is about exploring new, uncharted territories, not following a set path or reaching a specific endpoint. Several news articles have highlighted the dangers of this race metaphor,[8] describing how growing disagreements among powerful tech leaders have led to increased competition, pushing companies to accelerate advances in AI. This rush for dominance, as the article points out, shifts the focus from thoughtful, responsible development to a narrow pursuit of speed and competition, which can have harmful consequences for society.

Most important, if there were an AI race, it would be the wrong one. The current focus on building ever-larger models driven by vast amounts of data and computational power addresses only one aspect of intelligence. As we have seen in chapter 3, true intelligence involves reasoning, meaningful interaction, and joint decision-making in complex situations, which cannot be approximated by data-driven models alone. Moreover, as some studies suggest, the era of purely data-driven approaches may be reaching its limits, and we must explore alternative methods.[9] For example, recent research in weather forecasting and healthcare highlights the limitations of data-driven models, indicating a need for more integrated and comprehensive approaches.[10] Another study, by MIT,[11] analyzing twenty-five years of AI research, found that the dominance of deep learning, including generative AI, may be coming to an end.

It is more productive to view AI research as a collaborative exploration of a vast, uncharted landscape. When we focus too heavily on a single tracker, or a dominant approach, we lose the opportunity to invest in other promising paths.

Diversification in research and development is crucial. It allows us to explore different possibilities, compare methods, and choose approaches based on context and societal needs. Concentrating too many resources into one venture (such as pouring billions into OpenAI) limits our ability to address future challenges and prevents us from exploring alternative approaches that could be more effective in tackling urgent issues. In the long run, this lack of diversity in research can hinder innovation and leave us ill-prepared for the multifaceted problems AI will need to solve.

Moreover, and more important, AI that is opaque, untrustworthy, and unethical is unsustainable. The lack of transparency undermines trust, and the environmental toll of AI's resource use is a growing concern. The geopolitical competition for AI dominance exacerbates this "race to the bottom," prioritizing speed over ethics and sustainability, which could lead to global instability. The goal cannot be to win a race but rather to ensure that AI benefits humanity and the environment responsibly.

Safety above All?

In recent years, AI safety has emerged as the guiding principle of responsible AI development, particularly in international policy discussions. The emphasis on safety was evident in high-profile summits such as the U.K. AI Safety Summit at Bletchley Park (2023), the AI Seoul Summit (2024), and the AI Action Summit in Paris (2025), where world leaders, policymakers, and select technology companies convened to discuss the risks of advanced AI. These events have set the current tone for AI governance, promoting a framework where mitigating existential and technical risks takes precedence over broader societal concerns. Moreover, participation in these summits is limited to a select group of countries and a few strategic players, without clear criteria for inclusion. Many regions, particularly in the global South, are excluded from shaping the AI governance agenda, despite facing significant AI-related challenges. This raises concerns about whether the AI "safety-first" narrative is genuinely about global stability or merely serves to consolidate the influence of a few dominant powers.

The singular focus on AI safety reveals a dilemma: although safety is undoubtedly important, it cannot be the only measure of responsible AI development. AI is inherently a sociotechnical system; its risks and benefits extend beyond technical failures into societal, economic, and political domains. Yet, much of the discourse treats AI safety as a primarily technical challenge, solvable through improved alignment, monitoring, and regulation.[12] This narrow approach overlooks the complex tradeoffs AI presents: between human well-being and profit, openness and security, or efficiency and fairness.

The push for AI safety must be broadened beyond technical safeguards. A truly responsible approach requires a sociotechnical perspective—one that integrates ethical, economic, and geopolitical considerations. Safety is not an isolated issue; it is deeply intertwined with power structures, global inequalities, and economic priorities. Without a broader, more inclusive dialogue, AI governance risks being dictated by a small group of actors, prioritizing safety in ways that serve their interests while sidelining those of many others.

Where to Go from Here?

So, where to go from here? Concerns, such as those sparked in 2023 by the petition letter calling for a pause in AI development,[13] highlight the urgent need for regulatory frameworks to keep pace with AI advances. The pace at which new AI systems are released often outstrips the ability of legislators to create necessary legal frameworks, raising questions about whether all technologies should be developed and deployed, particularly when they may pose significant risks. Reflection is needed on the impact of these technologies, which are often introduced without clear user-driven demand. The development of AI is at a critical juncture, where the focus must shift from fear and speculation to responsible governance and ethical use. AI holds immense potential to transform industries and improve lives, but it also presents significant challenges, particularly when it comes to ensuring fairness, transparency, and accountability. We need algorithms that ensure accurate predictions while minimizing computational costs and energy use, models that are explainable, safe, and verifiable, and applications that benefit all rather than exploit vulnerable groups. These algorithms must handle causality, knowledge representation, context understanding, argumentation, negotiation, interaction, and abstraction.

We cannot accept the deployment of AI technologies without sufficient oversight, increasing the risk of bias, inequality, and other societal harms. To support oversight, social, legal, and organizational frameworks are essential. These frameworks should enforce accountability and liability, and acknowledge the limits of technology while providing safeguards for the outcomes of AI use. The focus should be as much on human development as on machine learning. These challenges are not purely engineering problems, nor can they be fully addressed by computational means alone—they demand innovative organizational processes, new educational curricula, and increased societal awareness of AI's possibilities and risks. AI, as

a scientific field, is inherently multidisciplinary, and true innovation necessitates embracing all disciplines, all groups, and all
differences.

Moreover, the concentration of AI power within a few large
tech companies raises concerns about monopolization and the
uneven distribution of benefits. It is essential to consider how
AI development is funded and controlled, and to explore the
need for publicly owned and controlled AI models to ensure
that these technologies serve the broader public interest rather
than the narrow goals of a few corporations. But in addition
to governance, there is a pressing need for multidisciplinary
research that combines technological expertise with insights
from the social sciences and humanities. This approach can
help address the complex ethical and societal challenges posed
by AI, ensuring that its development is aligned with human
values and societal needs.

Rather than viewing AI as a race to be won, it should be
seen as a collective endeavour that requires collaboration
across sectors and borders. The focus should be on creating
AI systems that are not only advanced but also safe, trustworthy, and beneficial for all. This requires moving beyond
techno-determinism—the belief that technology's course is
inevitable—and instead fostering a culture of informed public
oversight and democratic governance.

Ultimately, the goal should be to harness AI's potential in
ways that enhance human dignity, protect the environment,
and promote social justice. It is not just about avoiding risks but
about ensuring that AI technologies are developed and used in
ways that enhance human well-being and protect our planet.
This requires a shift from fear-driven narratives to proactive,
responsible action, ensuring that AI serves as a tool for positive
change rather than a source of division or harm.

Key Takeaways and Reflections

The solution paradox explores how solving problems with technology often creates new, unintended challenges. Although AI is often framed as the ultimate problem-solving tool, this techno-solutionist mindset risks oversimplifying complex societal issues and ignoring the broader consequences of technological interventions. AI's solutions often reflect the limitations of the data they are trained on, perpetuating existing biases and creating new inequities. For example, AI systems designed to optimize resource use may inadvertently widen social disparities by favoring efficiency over fairness. Additionally, overreliance on AI can lead to the neglect of alternative, human-centered approaches to addressing systemic challenges. Also, the growing focus on AI safety, emphasized in global summits and policy discussions, risks prioritizing technical fixes over broader socio-technical and geopolitical considerations.

True progress requires stepping back from the allure of purely technical fixes. Rather than relying solely on AI to provide solutions, we must consider the broader social, cultural, and ethical contexts in which these systems operate. Effective problem-solving combines technological innovation with critical reflection and human judgment, ensuring that solutions address the root causes rather than just solving the symptoms. The solution paradox reminds us to approach AI not as a magical answering tool. By recognizing its limitations and integrating diverse perspectives, we can harness AI's potential without losing sight of the values and complexities that define our shared challenges.

9

Epilogue

SHAPING TOMORROW

AS WE conclude our exploration of the paradoxes surrounding artificial intelligence, it is essential to reflect on the journey we have undertaken together. The AI paradox reminds us that despite the immense capabilities of AI, the core of human intelligence—our creativity, empathy, and moral reasoning—remains irreplaceable. AI, as powerful as it may become, is a reflection of human ingenuity, shaped by our choices, values, and the questions we dare to ask.

Throughout this book, I have argued that AI, as powerful as it may become, is not a monolithic force destined to replace us but rather a tool that amplifies our strengths and, at times, exposes our weaknesses. It challenges us to rethink what it means to be human in a world we increasingly share with intelligent machines. The paradoxes discussed, whether they relate to intelligence, justice, or power, serve as reminders that the future of AI is not predetermined. AI is not inevitable; it is what we, people, make of it. The future of AI will be determined by the decisions we make today, how we—not just developers and regulators, but all of us—design, influence, and apply these technologies, ensuring that everyone has a voice in shaping the direction AI takes. We must resist the seductive narratives that portray AI as an unstoppable force beyond human control, narratives that strip us of our agency and render us passive in the face of technological change. These narratives are dangerous because they absolve us of responsibility, fostering a sense of powerlessness in shaping the world around us.

The complexity of AI and the societal impacts it brings demand that we approach it with both humility and responsibility. Each of us has a role to play in this ongoing story. We must engage critically with the technology, question the motives behind its development, and actively participate in the discussions that will shape its future. We all must be vigilant, questioning the intentions behind the deployment of AI systems and the values they reflect. It is easy to fall into the trap of viewing AI as an inevitable force that dictates the terms of our existence. But remember that AI is a human creation and, as such, it can and should be guided by human principles and ethics. The stories we tell about AI, the ways we discuss it in public discourse, and the decisions we make in its development and regulation will determine whether it becomes a tool of empowerment or a source of control and inequality.

The paradoxes we face today are not the end of the story; they are the beginning of a new chapter in our collective journey. But that chapter is ours to write. The more we understand about AI, the more we will realize that the most profound questions it raises are not about technology but about ourselves. We must reject any narrative that seeks to render us powerless in the face of AI's advance. Instead, let us seize this moment with the determination to shape a future where AI enhances, rather than diminishes, the human experience. The power to decide lies with us—let us use it wisely and with unwavering vigilance.

In closing, the AI paradox is not a problem to be solved but a reality to be embraced. It challenges us to continuously reflect on our values, question our assumptions, and engage with the technology that is reshaping our world. As we look to the future, let us do so with a commitment to harnessing the potential while safeguarding the uniquely human qualities that define us. More than a paradox, the AI paradox is a call to action, a reminder that we are responsible for what we make of AI.

The more we rely on AI to shape our world, the more vital our human agency becomes in guiding its course.

NOTES

Chapter One

1. Jeffrey F. Cohn and Roger J. Kruez: Emotional intelligence in AI systems: Current challenges and future directions, in: *International Journal of Human-Computer Interaction* 35.12 (2019), pp. 1012–1030; Ameneh Shamekhi et al.: Understanding engagement in human-agent interaction: A mixed-methods investigation, in *Proceedings of the 2018 CHI Conference on Human Factors in Computing Systems (CHI '18)*, 2018. pp. 1–13.

2. Emily M. Bender et al.: On the dangers of stochastic parrots: Can language models be too big?, in: *Proceedings of the 2021 ACM Conference on Fairness, Accountability, and Transparency*, (2021), pp. 610–623, URL: https://doi.org/10.1145/3442188 .3445922.

3. Shannon Vallor: *The AI Mirror: How to Reclaim Our Humanity in an Age of Machine Thinking*, Oxford University Press 2024, URL: https://global.oup.com /academic/product/the-ai-mirror-9780197759066.

4. The OECD (Organization for Economic Cooperation and Development, https://www.oecd.org/) is a forum and knowledge hub for data, analysis, and best practices in public policy.

5. For a textbook definition of AI, see Stuart J. Russell and Peter Norvig: *Artificial Intelligence: A Modern Approach*, 4th ed., Pearson 2021.

6. See https://oecd.ai/en/wonk/definition for the rationale for the update.

7. Generative AI refers to systems capable of producing new content, such as text, images, or code, based on patterns in the data they are trained on, whereas LLMs are a specific type of AI designed to understand and generate humanlike text by leveraging vast amounts of data and advanced neural networks. See also: Roberto Gozalo-Brizuela and Eduardo C. Garrido-Merchan: ChatGPT is not all you need: A state of the art review of large generative AI models, in: arXiv preprint arXiv:2301.04655 2023, URL: https://arxiv.org/abs/2301.04655/.

8. Radiologists today balance caution and optimism regarding AI's role in their field, acknowledging its potential to enhance diagnostic accuracy while recognizing

that it cannot fully replace human expertise. See https://newrepublic.com/article
/187203/ai-radiology-geoffrey-hinton-nobel-prediction.

9. Shuroug A. Alowais et al.: Revolutionizing healthcare: The role of artificial
intelligence in clinical practice, in: *BMC Medical Education* 23.1 (2023), pp. 689.

10. Yu-Hao Li et al.: Innovation and challenges of artificial intelligence technology in personalized healthcare, in: *Scientific Reports* 14.1 (2024), pp. 18994.

11. See https://health.google/.

12. See https://www.pathai.com/.

13. Sreejan Kumar et al.: Disentangling abstraction from statistical pattern matching in human and machine learning, in: *PLoS Computational Biology* 19.8 (2023),
e1011316.

14. See https://bayes.cs.ucla.edu/WHY/.

15. Joshua B. Tenenbaum et al.: How to grow a mind: Statistics, structure, and
abstraction, in: *Science* 331.6022 (2011), pp. 1279–1285.

16. https://www.theguardian.com/culture/2023/oct/01/hollywood-writers
-strike-artificial-intelligence.

17. Artificial general intelligence (AGI) refers to AI systems that possess the
ability to understand, learn, and apply intelligence across a wide range of tasks at
a humanlike level, with the capacity to reason, solve problems, and adapt to new
situations autonomously.

18. Wicked problems are a well-known class of problems characterized by many
interconnected factors, incomplete information, and conflicting stakeholder interests, making them seem almost impossible to solve.

Chapter Two

1. For example, LG's ThinQ brand includes AI-driven home appliances. AI is also
used in services like personalized recommendations on platforms such as Netflix and
Amazon, and virtual assistants like Siri and Alexa.

2. Stephen Cave and Kanta Dihal: Hype, hope, and fear: The everyday AI
discourse and the shaping of collective imagination, in: *AI & Society* 34 (2019),
pp. 1–10.

3. Arvind Narayanan and Sayash Kapoor. *AI Snake Oil: What Artificial Intelligence
Can Do, What It Can't, and How to Tell the Difference*, Princeton University Press,
2024.

4. Ernesto Laclau and Chantal Mouffe: *Hegemony and Socialist Strategy: Towards
a Radical Democratic Politics* vol. 8. Verso Books 2014.

5. https://commission.europa.eu/system/files/2020-02/commission-white
-paper-artificial-intelligence-feb2020_en.pdf.

6. https://www.cnbc.com/2018/04/06/elon-musk-warns-ai-could-create-immortal-dictator-in-documentary.html.

7. https://www.thetimes.com/business-money/technology/article/ai-will-could-religions-to-to-control-humans-warns-sapiens-author-harari-fhbzgbv7b.

8. Andreas Theodorou and Virginia Dignum: Towards ethical and socio-legal governance in AI, in: *Nature Machine Intelligence* 2.1 (2020), pp. 10–12.

9. Melanie Mitchell: *Artificial Intelligence: A Guide for Thinking Humans*. Farrar, Straus & Giroux 2019. ISBN: 9780374257835.

10. Virginia Dignum: *Responsible Artificial Intelligence: How to Develop and Use AI in a Responsible Way*, Springer 2019.

11. See https://www.amnesty.org/en/documents/eur35/4686/2021/en/.

12. See https://www.bbc.com/news/world-europe-55674146.

13. Anna Jobin et al.: The global landscape of AI ethics guidelines, in: *Nature Machine Intelligence* 1 (2019), pp. 389–399.

14. Kate Crawford: *The Atlas of AI: Power, Politics, and the Planetary Costs of Artificial Intelligence*, Yale University Press 2021.

15. For further details on these applications, I suggest the following articles, including work I did with colleagues on the societal impact of COVID-19 policies. Irene Aldridge: High-Frequency Trading: A Practical Guide to Algorithmic Strategies and Trading Systems, John Wiley & Sons 2013; Stephen Burgess et al.: Agent-based models: Understanding the economy from the bottom up, in: *Bank of England Quarterly Bulletin* 56.3 (2016), pp. 178–190; Frank Dignum et al.: Analysing the combined health, social and economic impacts of the coronavirus pandemic using agent-based social simulation, in: arXiv preprint arXiv:2004.12809 2020.

16. For example, watch this talk: https://www.youtube.com/watch?v=IsMY6N ZZnWc.

17. Similar themes have been explored by others. For example, David Gunkel, a prominent scholar in the fields of AI ethics and philosophy of technology, explores the ethical implications of considering AI as moral agents or patients, questioning whether machines with humanlike behavior deserve moral status and examining the challenges of defining machine agency. Leading AI figure Stuart Russell discusses the risks of autonomous AI that lacks alignment with human values, stressing the need for provably beneficial AI systems that adhere to ethical principles. David J. Gunkel: *The Machine Question: Critical Perspectives on AI, Robots, and Ethics*. MIT Press 2012, p. 5; Stuart Russell: *Human Compatible: AI and the Problem of Control*. Penguin 2019.

18. The perspective of viewing AI as a tool, focusing on its practical applications rather than replicating human traits, has been discussed in various works. In "The Turing trap," economist Erik Brynjolfsson argues that augmenting human labor with AI tools can lead to greater economic and social benefits. Similarly, the Vatican's 2025

guidelines on AI ethics underscore that AI should complement rather than replace human intelligence. Erik Brynjolfsson: The Turing trap: The promise and peril of humanlike artificial intelligence, in: arXiv preprint arXiv:2201.04200 (2022), URL: https://arxiv.org/abs/2201.04200.

19. Reflections on the implications of the different perspectives include work by Bender and colleagues, who report that AI simulations aid prediction but rely on data accuracy and bias, whereas Mittelstadt and colleagues identify that direct applications raise higher stakes, requiring clear accountability. Emily M. Bender et al.: On the dangers of stochastic parrots: Can language models be too big? *Proceedings of the 2021 ACM Conference on Fairness, Accountability, and Transparency* 2021, pp. 610–623. URL: https://doi.org/10.1145/3442188.3445922; Brent Mittelstadt et al.: The ethics of algorithms: Mapping the debate, in: *Big Data & Society* 3.2 (2016), URL: https://doi.org/10.1177/2053951716679679.

Chapter Three

1. See Kate Crawford: *The Atlas of AI: Power, Politics, and the Planetary Costs of Artificial Intelligence*, Yale University Press 2021.

2. Ada Lovelace: Sketch of the analytical engine invented by Charles Babbage, Esq., in: *Scientific Memoirs* 3 (1843): pp. 666–731.

3. See https://en.wikipedia.org/wiki/Larry_Tesler.

4. See Dunbar, R. I. The social brain hypothesis and its implications for social evolution. *Annals of Human Biology*, 36 n.5 (2009), 562–572.

5. Hesheng Liu et al.: Evidence from intrinsic activity that asymmetry of the human brain is controlled by multiple factors, in: *Proceedings of the National Academy of Sciences* 106.48 (2009), pp. 20499–20503.

6. G. Buzsáki: *Rhythms of the Brain*, Oxford University Press 2011.

7. J. R. Searle: *The Rediscovery of the Mind*, MIT Press 1992.

8. Created by Rudolph Zallinger for the 1965 book *Early Man*, by F. Clark Howell.

9. A theoretical model of computation (that is, not physically realizable) that, at each step, can branch into multiple possible states simultaneously, as if exploring all computation paths at once.

10. NP (nondeterministic polynomial time) is a class of problems where, if given a possible solution, a computer can quickly verify its correctness, even though finding the solution from scratch might take an extremely long time. For example, solving a Sudoku puzzle is hard, but once someone gives you a completed solution, it's easy to check if it's correct.

11. See also Iris van Rooij et al.: Reclaiming ai as a theoretical tool for cognitive science, in: *Computational Brain & Behavior* 7 (2024): pp. 616–636, URL: https://doi .org/10.31234/osf.io/4cbuv.

12. https://en.wikipedia.org/wiki/Moravec's_paradox.

13. https://www.talend.com/blog/the-ai-paradox-and-the-importance-of -human-led-automated-intelligence/.

14. Virginia Dignum: *Responsible Artificial Intelligence: How to Develop and Use AI in a Responsible Way*, Springer 2019.

15. Stuart J. Russell and Peter Norvig: *Artificial Intelligence: A Modern Approach*, 4th ed., Pearson 2021.

16. Virginia Dignum: Social agents: Bridging simulation and engineering, in: *Communications of the ACM* 60.11 (2017), pp. 32–34.

17. Elizabeth Gibney: AI models fed AI-generated data quickly spew nonsense, in: *Nature* 632.8023 (Aug. 2024), pp. 18–19, URL: https://ideas.repec.org/a/nat /nature/v632y2024i8023d10.1038_d41586-024-02420-7.html.

18. Boris van Breugel and Mihaela van der Schaar: Beyond privacy: Navigating the opportunities and challenges of synthetic data, 2023, URL: https://arxiv.org/abs /2304.03722.

19. James Jordon et al.: Synthetic data—What, why and how?, 2022, URL: https://arxiv.org/abs/2205.03257.

20. A good visualization of this is available in the Mozilla Foundation Internet health report of 2022; see https://2022.internethealthreport.org/.

21. https://garymarcus.substack.com/p/this-one-important-fact-about-current

22. In philosophy, *bullshit* is characterized by a disregard for truth, making it more dangerous to society than lying because it erodes the value of truth in communication. See Harry G. Frankfurt: *On Bullshit*, Princeton University Press 2005. Originally published as an essay in 1986.

23. Michael Townsen Hicks, James Humphries, and Joe Slater: Chatgpt is bullshit, in: *Ethics and Information Technology* 26 (2024), p. 38, URL: https://doi.org/10 .1007/s10676-024-09775-5.

24. Angela D. Friederici: The brain basis of language processing: From structure to function, in: *Physiological Reviews* 91.4 (2011), pp. 1357–1392; Zoha Deldar et al.: The interaction between language and working memory: A systematic review of fMRI studies in the past two decades, in: *AIMS Neuroscience* 8.1 (2021), p. 1.

Chapter Four

1. Atoosa Kasirzadeh: Algorithmic fairness and structural injustice: Insights from feminist political philosophy, in: *Proceedings of the 2022 AAAI/ACM Conference on AI, Ethics, and Society*, 2022, pp. 349–356.

2. Bruce Schneier: *Secrets and Lies: Digital Security in a Networked World*, John Wiley & Sons 2000.

3. John Rawls: *A Theory of Justice*. Belknap Press 1971.

4. In ethical philosophy, utilitarianism is a family of normative ethical theories that prescribe actions that maximize happiness and well-being for the affected individuals. In other words, utilitarian ideas encourage actions that ensure the greatest good for the greatest number. [Wikipedia, https://en.wikipedia.org/wiki/Utilitarianism].

5. E.g., FAccT: https://facctconference.org/.

6. One well-known example is Amazon's AI recruiting tool, which was scrapped in 2018 after it was found to be biased against women. The machine-learning algorithm, trained on résumés submitted over a ten-year period (which were predominantly from men), learned to favor male candidates and penalized résumés that included terms like *women's* (e.g., "women's chess club") or came from all-women's colleges, reinforcing gender bias instead of eliminating it.

7. Brian Hsu et al.: Pushing the limits of fairness impossibility: Who's the fairest of them all?, in: *Advances in Neural Information Processing Systems* 35 (2022), pp. 32749–32761; Sorelle A. Friedler, Carlos Scheidegger, and Suresh Venkatasubramanian: The (im)possibility of fairness: Different value systems require different mechanisms for fair decision making, in: *Communications of the ACM* 64.4 (2021), pp. 136–143.

8. See also Brian Christian's *The Alignment Problem* (2020), which traces the evolution of technical work on fairness and interpretability and shows how these challenges are inseparable from broader societal and ethical concerns.

9. This example is at the core of the well-known ProPublica investigations of the COMPAS algorithms used by courts in the United States to determine recidivism risk: www.propublica.org/article/how-we-analyzed-the-compasrecidivism-algorithm.

10. https://aif360.res.ibm.com/.

11. https://research.google/pubs/the-what-if-tool-interactive-probing-of-machine-learning-models/.

12. David Sumpter: *Outnumbered: From Facebook and Google to Fake News and Filter-Bubbles–The Algorithms That Control Our Lives*. Bloomsbury Publishing 2018.

13. Virginia Dignum: *Responsible Artificial Intelligence: How to Develop and Use AI in a Responsible Way*, Springer 2019.

Chapter Five

1. https://www.progressivepolicy.org/publication/a-brief-history-of-internet-regulation-2/.

2. Jessica Fjeld et al.: Governance of artificial intelligence, in: *Harvard Journal of Law & Technology* (2020).

3. Anne Jobin et al.: The global landscape of AI ethics guidelines, in: *Nature Machine Intelligence* 1 (2019): pp. 389–399.

4. Ryan Calo: Artificial intelligence policy: A primer and roadmap, in: *University of California Davis Law Review* 51 (2017), URL: https://digitalcommons.law.uw.edu /faculty-articles/640.

5. Jack M. Balkin: Fixing social media's grand bargain, Hoover Working Group on National Security, Technology, and Law, Aegis series paper no. 1814, October 16, 2018, SSRN, URL: https://papers.ssrn.com/sol3/papers.cfm?abstract_id =3266942.

6. Stuart Russell: *Human Compatible: AI and the Problem of Control*, Penguin 2019.

7. Daniel Araya: Regulating artificial intelligence: The challenge of incentives, in: *Technology Policy* (2020).

8. Lydia Nicholas: Impact assessment in AI regulation, in: *Journal of Responsible Technology* (2020).

9. Matthias Spielkamp: Accountability frameworks for algorithmic systems, in: *AlgorithmWatch* (2021).

10. https://www.un.org/en/ai-advisory-body.

11. Fatality rate, or mortality rate, measures the number of deaths in a particular population, scaled to the size of that population, per unit of time.

12. See https://blogs.lse.ac.uk/covid19/2021/04/21/the-nhs-contact-tracing -app-fell-foul-of-privacy-concerns-but-did-they-have-the-right-idea/.

13. See https://www.ft.com/content/1d468bd2-6712-4cdd-ac71-21e0ace2d048 and https://e360.yale.edu/features/artificial-intelligence-climate-energy-emissions.

14. https://www.nber.org/reporter/fall-2006/economics-pharmaceutical -industry.

15. https://www.fca.org.uk/news/speeches/how-innovation-and-regulation-in -financial-services-can-drive-uk-economic-growth.

16. https://www.mdpi.com/1660-4601/19/23/16290.

17. https://www.energy.gov/eere/office-energy-efficiency-renewable-energy

18. https://www.deloitte.com/global/en/our-thinking/insights/industry/gover nment-public-services/government-trends/2023/regulatory-agencies-and-innova- tion.html.

19. https://www.itu.int/en/ITU-D/Regulatory-Market/Documents/Events20 21/Impact_policies_regulation_and_Institutions_on_ICT_Sector_performance .pdf.

20. Philippe Aghion, Antonin Bergeaud, and John Van Reenen: The impact of regulation on innovation, in: *American Economic Review* 113.11 (2023), pp. 2894–2936.

21. https://commission.europa.eu/sites/default/files/commission-white-paper -artificial-intelligence-feb2020_en.pdf.

22. AI Now Institute: Algorithmic impact assessments: A practical framework for public agency accountability (2018). URL: https://ainowinstitute.org/aiareport2018 .pdf.

Chapter Six

1. See Tim Urban's *What's Our Problem*, Claymont, DE 2023. https://waitbutwhy .com/whatsourproblem.

2. See https://futureoflife.org/document/fli-ai-safety-index-2024/.

3. See https://www.theguardian.com/technology/2017/may/16/google-deep mind-16m-patient-record-deal-inappropriate-data-guardian-royal-free.

4. See https://www.theguardian.com/technology/2025/jan/10/mark-zucker berg-meta-factchecking.

5. See https://www.technologyreview.com/2020/02/17/844721/ai-openai -moonshot-elon-musk-sam-altman-greg-brockman-messy-secretive-reality/.

6. See https://authorsguild.org/news/ag-and-authors-file-class-action-suit -against-openai.

7. See https://www.saverilawfirm.com/our-cases/github-copilot-intellectual -property-litigation.

8. See https://www.theguardian.com/technology/2024/sep/27/openai-shift -to-for-profit-company-may-lead-it-to-cut-corners-says-whistleblower.

9. See https://www.wired.com/story/openai-voice-mode-emotional-attach ment.

10. See https://www.vox.com/future-perfect/367188/love-addicted-ai-voice -human-gpt4-emotion.

11. See https://www.prnewswire.com/news-releases/concerns-around-ethi cal-risks-of-generative-ai-remain-high-amid-increased-adoption-deloitte-state-of -ethics-and-trust-in-technology-report-302254593.html.

12. Mark Haugaard: The four dimensions of power: Conflict and democracy, in: *Journal of Political Power* 14.1 (2021), pp. 153–175, URL: https://www.tandfonline.com /doi/full/10.1080/2158379X.2021.1878411.

13. John R.P. French and Bertram Raven: The bases of social power, in: Dorwin Cartwright (ed.): *Studies in Social Power*, Ann Arbor, MI 1959, pp. 150–167.

14. https://thealliance.ai/.

15. Tim Miller: Explanation in artificial intelligence: Insights from the social sciences, in: *Artificial Intelligence* 267 (2019), pp. 1–38.

16. https://www.thebureauinvestigates.com/stories/2024-03-27/how-repres sive-regimes-use-facial-recognition-technology/.

17. See https://jpia.princeton.edu/news/social-credit-system-not-just-another-chinese-idiosyncrasy.

18. See https://www.bbc.com/news/technology-57101248.

19. https://news.mit.edu/2023/how-ai-tocracy-emerges-0713.

20. https://www.hrw.org/news/2021/09/15/russia-broad-facial-recognition-use-undermines-rights.

21. https://www.cfr.org/blog/ai-assault-women-what-irans-tech-enabled-morality-laws-indicate-womens-rights-movements.

22. https://www.newarab.com/analysis/how-ai-big-tech-and-spyware-power-israels-occupation.

23. https://securityandtechnology.org/blog/decrypting-irans-ai-enhanced-operations-in-cyberspace/.

24. See https://carnegieendowment.org/research/2019/09/the-global-expansion-of-ai-surveillance?lang=en, https://www.wfd.org/what-we-do/resources/democratic-approach-global-ai-safety/how-ai-might-impact-democracy and https://thebulletin.org/2024/06/how-ai-surveillance-threatens-democracy-everywhere/.

25. https://www.atlanticcouncil.org/blogs/geotech-cues/the-west-china-and-ai-surveillance/.

26. See https://www.euractiv.com/section/global-europe/news/russian-ai-generated-propaganda-to-pose-more-problems-for-ukraine/.

27. See the Human Rights Watch report "China's Algorithms of Repression," https://www.hrw.org/report/2019/05/01/chinas-algorithms-repression/reverse-engineering-xinjiang-police-mass. For an overview of AI related human rights violations and concerns, see https://www.europarl.europa.eu/RegData/etudes/IDAN/2024/754450/EXPO_IDA(2024)754450_EN.pdf.

28. See https://www.computerweekly.com/news/366599014/Ban-predictive-policing-and-facial-recognition-says-civil-society.

29. See AAAI 2025 Presidential Panel on the Future of AI Research, Association for the Advancement of Artificial Intelligence, https://aaai.org/wp-content/uploads/2025/03/AAAI-2025-PresPanel-Report-Digital-3.7.25.pdf.

30. Daron Acemoglu and Pascual Restrepo: Automation and new tasks: How technology displaces and reinstates labor, in: *Journal of Economic Perspectives* 33.2 (2019), pp. 3–30.

31. See https://www.pbs.org/newshour/economy/smart-robots-will-take-third-jobs-2025-gartner-says.

32. See https://www.gartner.com/en/newsroom/press-releases/2017-12-13-gartner-says-by-2020-artificial-intelligence-will-create-more-jobs-than-it-eliminates.

33. See https://www.gartner.com/en/documents/4526399.

Chapter Seven

1. Johannes Jaeger: Artificial intelligence is algorithmic mimicry: Why artificial "agents" are not (and won't be) proper agents, in: *Neurons, Behavior, Data Analysis, and Theory* 2024, URL: http://dx.doi.org/10.51628/001c.94404; and Emily M. Bender et al.: On the dangers of stochastic parrots: Can language models be too big?, in: *Proceedings of the 2021 ACM Conference on Fairness, Accountability, and Transparency* 2021, pp. 610–628. URL: https://doi.org/10.1145/3442188.3445922.

2. See https://www.schneier.com/books/secrets-and-lies-pref.

3. See https://www.businessinsider.com/eric-schmidt-google-ai-data-centers-energy-climate-goals-2024-10.

4. We will further discuss this issue of techno-solutionism in the next chapter.

5. Read this essay by François Chollet https://www.freethink.com/robots-ai/arc-prize-agi, or listen to Gary Marcus in https://podcasts.apple.com/za/podcast/gary-marcus-keynote-at-agi-24/id1510472996?i=1000665733355.

6. Gary Marcus: The next decade in AI: Four steps towards robust artificial intelligence, in: arXiv preprint arXiv:2002.06177 2020, URL: https://arxiv.org/abs/2002.06177.

7. Laura Weidinger et al.: Ethical and social risks of harm from language models, 2021. URL: https://arxiv.org/abs/2112.04359.

8. Daniel C. Dennett: Intentional systems theory, in: Brian McLaughlin, Ansgar Beckermann, and Sven Walter (eds.) *The Oxford Handbook of Philosophy of Mind*, Oxford University Press 2009, pp. 339–350.

9. Murray Shanahan: Talking about large language models, 2023, URL: https://arxiv.org/abs/2212.03551.

10. An intractable, or NP-hard, problem is one that is extremely difficult to solve quickly, and no efficient computational solution is known. The bigger the problem gets, the more the required time and resources grow exponentially, making it nearly impossible to solve quickly for large instances.

11. Iris van Rooij et al.: Reclaiming AI as a theoretical tool for cognitive science, in: *Computational Brain & Behavior* 7 (2024): pp. 616–636.

12. See chapter 3.

13. Yoshua Bengio et al.: Managing extreme ai risks amid rapid progress, in: *Science* 384.6698 (2024), pp. 842–845. URL: https://www.science.org/doi/abs/10.1126/science.adn0117.

14. Remember the quotation by Bruce Schneier earlier in this chapter.

15. Already in 1863, Samuel Butler, in his essay "Darwin among the Machines," warned: *"Day by day, however, the machines are gaining ground upon us; day by day we are becoming more subservient to them; more men are daily bound down as slaves to tend them."*

16. See https://en.wikipedia.org/wiki/Eugene_Goostman.

17. Nick Bostrom: *Superintelligence: Paths, Dangers, Strategies*, Oxford University Press 2014.

18. Ziad Obermeyer et al.: Dissecting racial bias in an algorithm used to manage the health of populations, in: *Science* 366.6464 (2019), pp. 447–453. URL: https://www.science.org/doi/10.1126/science.aax2342.

19. See https://www.wri.org/insights/interactive-chart-shows-changes-worlds -top-10-emitters.

20. Bengio et al.: Managing extreme ai risks amid rapid progress.

21. Zeynep Akata et al.: A research agenda for hybrid intelligence: Augmenting human intellect by collaborative, adaptive, responsible and explainable artificial intelligence, in: *Computer* 53.8 (2020), pp. 18–28.

Chapter Eight

1. Simon Lindgren and Virginia Dignum: Beyond AI solutionism: Toward a multi-disciplinary approach to artificial intelligence in society, in: *Handbook of Critical Studies of Artificial Intelligence*, Edward Elgar Publishing 2023, pp. 163–172.

2. Panagiota Galetsi, Korina Katsaliaki, and Sameer Kumar: The medical and societal impact of big data analytics and artificial intelligence applications in combating pandemics: A review focused on covid-19, in: *Social Science & Medicine* 301 (2022), pp. 114973. URL: https://www.sciencedirect.com/science/article/pii /S0277953622002799.

3. See https://cacm.acm.org/blogcacm/the-software-that-led-to-the-lock down.

4. https://www.deepseek.com/

5. https://foreignpolicy.com/2025/02/05/deep-seek-china-us-artificial-intellige nce-ai-arms-race/

6. https://gpai.ai/

7. https://www.un.org/en/ai-advisory-body

8. See, for example, https://www.nytimes.com/2023/12/08/briefing/ai-domi nance.html.

9. Zied Ben-Bouallegue et al.: The rise of data-driven weather forecasting, in: arXiv preprint arXiv:2307.10128 2023, Accessed: 2024-06-08, URL: https://arxiv.org /abs/2307.10128.

10. BMC Health Services Research: Learning health systems using data to drive healthcare improvement and impact: A systematic review, in: *BMC Health Services Research* 2023, Accessed: 2024-06-08, URL: https://bmchealthservres .biomedcentral.com/articles/10.1186/s12913-020-05856-1.

11. https://www.technologyreview.com/2023/01/19/1069605/we-analyzed-166 25-papers-to-figure-out-where-ai-is-headed-next/.

12. See, for example, the recent report on AI safety: Yoshua Bengio et al.: International AI safety report, 2025, URL: https://arxiv.org/abs/2501.17805.

13. See https://futureoflife.org/open-letter/pause-giant-ai-experiments.

BIBLIOGRAPHY

Acemoglu, Daron, and Pascual Restrepo: Automation and new tasks: How technology displaces and reinstates labor, in: *Journal of Economic Perspectives* 33.2 (2019), pp. 3–30.

Aghion, Philippe, Antonin Bergeaud and John Van Reenen: The impact of regulation on innovation, in: *American Economic Review* 113.11 (2023), pp. 2894–2936.

AI Now Institute: Algorithmic impact assessments: A practical framework for public agency accountability, (2018), URL: https://ainowinstitute.org/aiareport2018 .pdf.

Akata, Zeynep, et al.: A research agenda for hybrid intelligence: Augmenting human intellect by collaborative, adaptive, responsible and explainable artificial intelligence, in: *Computer* 53.8 (2020), pp. 18–28.

Aldridge, Irene: *High-Frequency Trading: A Practical Guide to Algorithmic Strategies and Trading Systems*, John Wiley & Sons (2013).

Alowais, Shuroug A., et al: Revolutionizing healthcare: the role of artificial intelligence in clinical practice, in: *BMC Medical Education* 23.1 (2023), p. 689.

Araya, Daniel: Regulating artificial intelligence: The challenge of incentives, in: *Technology Policy* (2020).

Balkin, Jack M.: Fixing social media's grand bargain, Hoover Working Group on National Security, Technology, and Law, Aegis Series Paper no. 1814, October 16, 2018, SSRN, URL: https://papers.ssrn.com/sol3/papers.cfm?abstract _id=3266942.

Ben-Bouallegue, Zied, et al.: The rise of data-driven weather forecasting, in: arXiv preprint arXiv:2307.10128 [physics-ao] (2023), Accessed: 2024-06-08, URL: https://arxiv.org/abs/2307.10128.

Bender, Emily M., et al.: On the dangers of stochastic parrots: Can language models be too big?, in: *Proceedings of the 2021 ACM Conference on Fairness, Accountability, and Transparency* (2021), pp. 610–623. URL: https://doi.org/10.1145/3442188 .3445922.

Bengio, Yoshua, et al.: International AI safety report, arXiv:2501.17805 [cs.CY] 2025. URL: https://arxiv.org/abs/2501.17805.

Bengio, Yoshua, et al.: Managing extreme AI risks amid rapid progress, in: *Science* 384.6698 (2024), pp. 842–845. URL: https://www.science.org/doi/abs/10.1126/science.adn0117.

BMC Health Services Research: Learning health systems using data to drive healthcare improvement and impact: A systematic review, in: *BMC Health Services Research* 2023, Accessed: 2024-06-08, URL: https://bmchealthservres.biomedcentral.com/articles/10.1186/s12913-020-05856-1.

Bostrom, Nick: *Superintelligence: Paths, Dangers, Strategies*, Oxford University Press 2014.

Breugel, Boris van, and Mihaela van der Schaar: Beyond privacy: Navigating the opportunities and challenges of synthetic data, 2023, URL: https://arxiv.org/abs/2304.03722.

Brynjolfsson, Erik: The Turing trap: The promise & peril of humanlike artificial intelligence, in: arXiv preprint arXiv:2201.04200 2022, URL: https://arxiv.org/abs/2201.04200.

Burgess, Stephen, et al.: Agent-based models: Understanding the economy from the bottom up, in: *Bank of England Quarterly Bulletin* 56.3 (2016), pp. 178–190.

Buzsáki, G.: *Rhythms of the Brain*, Oxford 2011.

Calo, Ryan: Artificial intelligence policy: A primer and roadmap, in: *University of California Davis Law Review* 51 (2017), URL: https://digitalcommons.law.uw.edu/faculty-articles/640.

Cave, Stephen, and Kanta Dihal: Hype, hope, and fear: The everyday AI discourse and the shaping of collective imagination, in: *AI & Society* 34 (2019), pp. 1–10.

Christian, B.: *The Alignment Problem: Machine Learning and Human Values*, W. W. Norton & Company 2020.

Cohn, Jeffrey F., and Roger J. Kruez: Emotional intelligence in AI systems: Current challenges and future directions, in: *International Journal of Human-Computer Interaction* 35.12 (2019), pp. 1012–1030.

Crawford, Kate: *The Atlas of AI: Power, Politics, and the Planetary Costs of Artificial Intelligence*, Yale University Press 2021.

Deldar, Zoha, et al.: The interaction between language and working memory: A systematic review of fMRI studies in the past two decades, in: *AIMS Neuroscience* 8.1 (2021), p. 1.

Dennett, Daniel C.: Intentional systems theory, in, Brian McLaughlin, Ansgar Beckermann, and Sven Walter (eds.): *The Oxford Handbook of Philosophy of Mind*, Oxford University Press 2009, pp. 339–350.

Dignum, Frank, et al.: Analysing the combined health, social and economic impacts of the coronavirus pandemic using agent-based social simulation, in: arXiv preprint arXiv:2004.12809 2020.

Dignum, Virginia: *Responsible Artificial Intelligence: How to Develop and Use AI in a Responsible Way*, Springer 2019.

Dignum, Virginia: Social agents: Bridging simulation and engineering, in: *Communications of the ACM* 60.11 (Nov. 2017), pp. 32–34.

Domingos, Pedro: *The Master Algorithm: How the Quest for the Ultimate Learning Machine Will Remake Our World*, New York 2015. URL: https://www.amazon.com/The-Master-Algorithm-Ultimate-Learning/dp/0062409171.

Fjeld, Jessica, et al.: Governance of artificial intelligence, in: *Harvard Journal of Law & Technology* (2020).

Frankfurt, Harry G.: *On Bullshit*, Princeton, NJ 2005. Originally published as an essay in 1986.

French, John R.P., and Bertram Raven: The bases of social power, in Dorwin Cartwright (ed.): *Studies in Social Power*, Ann Arbor, MI 1959, pp. 150–167.

Friederici, Angela D.: The brain basis of language processing: From structure to function, in: *Physiological Reviews* 91.4 (2011), pp. 1357–1392.

Friedler, Sorelle A., Carlos Scheidegger and Suresh Venkatasubramanian: The (im)possibility of fairness: Different value systems require different mechanisms for fair decision making, in: *Communications of the ACM* 64.4 (2021), pp. 136–143.

Galetsi, Panagiota, Korina Katsaliaki, and Sameer Kumar: The medical and societal impact of big data analytics and artificial intelligence applications in combating pandemics: A review focused on Covid-19, in: *Social Science & Medicine* 301 (2022), p. 114973, URL: https://www.sciencedirect.com/science/article/pii/S0277953622002799.

Gibney, Elizabeth: AI models fed AI-generated data quickly spew nonsense, in: *Nature* 632.8023 (Aug. 2024), pp. 18–19. URL: https://ideas.repec.org/a/nat/nature/v632y2024i8023d10.1038_d41586-024-02420-7.html.

Gozalo-Brizuela, Roberto, and Eduardo C. Garrido-Merchan: ChatGPT is not all you need: A state of the art review of large generative AI models, in: arXiv preprint arXiv:2301.04655 (2023), URL: https://arxiv.org/abs/2301.04655.

Gray, Mary L., and Siddharth Suri: *Ghost Work: How to Stop Silicon Valley from Building a New Global Underclass*, Boston 2019.

Gunkel, David J.: *The Machine Question: Critical Perspectives on AI, Robots, and Ethics*, MIT Press 2012.

Haugaard, Mark: The four dimensions of power: Conflict and democracy, in: *Journal of Political Power* 14.1 (2021), pp. 153–175. URL: https://www.tandfonline.com/doi/full/10.1080/2158379X.2021.1878411.

Hicks, Michael Townsen, James Humphries, and Joe Slater: ChatGPT is bullshit, in: *Ethics and Information Technology* 26 (2024), p. 38. URL: https://doi.org/10.1007/s10676-024-09775-5.

Hsu, Brian, et al.: Pushing the limits of fairness impossibility: Who's the fairest of them all?, in: *Advances in Neural Information Processing Systems* 35 (2022), pp. 32749–32761.

Howell, F. Clark: *Early Man*, New York 1965.

Jaeger, Johannes: Artificial intelligence is algorithmic mimicry: Why artificial "agents" are not (and won't be) proper agents, in: *Neurons, Behavior, Data Analysis, and Theory* 2024, URL: http://dx.doi.org/10.51628/001c.94404.

Jobin, Anna, et al.: The global landscape of AI ethics guidelines, in: *Nature Machine Intelligence* 1 (2019), pp. 389–399.

Jordon, James, et al.: Synthetic data—What, why and how?, 2022, URL: https://arxiv.org/abs/2205.03257.

Kasirzadeh, Atoosa: Algorithmic fairness and structural injustice: Insights from feminist political philosophy, in *Proceedings of the 2022 AAAI/ACM Conference on AI, Ethics, and Society*, 2022, pp. 349–356.

Kumar, Sreejan, et al.: Disentangling abstraction from statistical pattern matching in human and machine learning, in: *PLoS Computational Biology* 19.8 (2023), e1011316.

Laclau, Ernesto, and Chantal Mouffe: *Hegemony and Socialist Strategy: Towards a Radical Democratic Politics*, vol 8, Verso Books 2014.

Li, Yu-Hao, et al.: Innovation and challenges of artificial intelligence technology in personalized healthcare, in: *Scientific Reports* 14.1 (2024), p. 18994.

Lindgren, Simon, and Virginia Dignum: Beyond AI solutionism: Toward a multidisciplinary approach to artificial intelligence in society, in *Handbook of Critical Studies of Artificial Intelligence*, Edward Elgar Publishing 2023, pp. 163–172.

Liu, Hesheng, et al.: Evidence from intrinsic activity that asymmetry of the human brain is controlled by multiple factors, in: *Proceedings of the National Academy of Sciences* 106.48 (2009), pp. 20499–20503.

Lovelace, Ada: Sketch of the analytical engine invented by Charles Babbage, Esq., in: *Scientific Memoirs* 3 (1843), pp. 666–731.

Marcus, Gary: The next decade in AI: Four steps towards robust artificial intelligence, in: arXiv preprint arXiv:2002.06177 (2020), URL: https://arxiv.org/abs/2002.06177.

Miller, Tim: Explanation in artificial intelligence: Insights from the social sciences, in: *Artificial Intelligence* 267 (2019), pp. 1–38.

Mitchell, Melanie: *Artificial Intelligence: A Guide for Thinking Humans*, Farrar, Straus & Giroux 2019.

Mittelstadt, Brent, et al.: The ethics of algorithms: Mapping the debate, in: *Big Data & Society* 3.2 (2016), URL: https://doi.org/10.1177/2053951716679679.

Narayanan, Arvind, and Sayash Kapoor: *AI Snake Oil: What Artificial Intelligence Can Do, What It Can't, and How to Tell the Difference*, Princeton University Press 2024.

Nicholas, Lydia: Impact assessment in AI regulation, in: *Journal of Responsible Technology* (2020).

Obermeyer, Ziad, et al.: Dissecting racial bias in an algorithm used to manage the health of populations, in: *Science* 366.6464 (2019), pp. 447–453. URL: https://www.science.org/doi/10.1126/science.aax2342.

Rawls, John: *A Theory of Justice*. Belknap Press 1971.

Rooij, Iris van, et al.: Reclaiming AI as a theoretical tool for cognitive science, in: *Computational Brain & Behavior* 7 (2024), pp. 616–636, URL: https://doi.org/10.31234/osf.io/4cbuv.

Russell, Stuart: *Human Compatible: AI and the Problem of Control*, Penguin 2019.

Russell, Stuart J., and Peter Norvig: *Artificial Intelligence: A Modern Approach*, 4th ed., Pearson 2021.

Schneier, Bruce: *Secrets and Lies: Digital Security in a Networked World*. John Wiley & Sons 2000.

Searle, J. R.: *The Rediscovery of the Mind*, MIT Press 1992.

Shamekhi, Ameneh, et al.: Understanding engagement in human-agent interaction: A mixed-methods investigation, in *Proceedings of the 2018 CHI Conference on Human Factors in Computing Systems (CHI '18)* 2018, pp. 1–13.

Shanahan, Murray: Talking about large language models, 2023, URL: https://arxiv.org/abs/2212.03551.

Spielkamp, Matthias: Accountability frameworks for algorithmic systems, in: *AlgorithmWatch* (2021).

Sumpter, David: *Outnumbered: From Facebook and Google to Fake News and Filter-Bubbles–the Algorithms that Control Our Lives*, Bloomsbury Publishing 2018.

Tenenbaum, Joshua B., et al.: How to grow a mind: Statistics, structure, and abstraction, in: *Science* 331.6022 (2011), pp. 1279–1285.

Theodorou, Andreas, and Virginia Dignum: Towards ethical and socio-legal governance in AI, in: *Nature Machine Intelligence* 2.1 (2020), pp. 10–12.

Urban, Tim: *What's Our Problem*, Claymont, DE 2023.

Vallor, Shannon: *The AI Mirror: How to Reclaim Our Humanity in an Age of Machine Thinking*. Oxford University Press 2024, URL: https://global.oup.com/academic/product/the-ai-mirror-9780197759066.

Weidinger, Laura, et al.: Ethical and social risks of harm from language models, 2021, URL: https://arxiv.org/abs/2112.04359.

INDEX

GPSR Authorized Representative: Easy Access System Europe - Mustamäe tee
50, 10621 Tallinn, Estonia, gpsr.requests@easproject.com

www.ingramcontent.com/pod-product-compliance
Lightning Source LLC
Chambersburg PA
CBHW051728260326
41914CB00040B/2006/J